# my revision notes

## Edexcel International GCSE (9–1)

# PHYSICS

Ian Horsewell

HODDER
EDUCATION
AN HACHETTE UK COMPANY

The publisher would like to thank the following for permission to reproduce copyright material:

**Photo credits:** p.74 © Fotimmz – Fotolia

Although every effort has been made to ensure that website addresses are correct at time of going to press, Hodder Education cannot be held responsible for the content of any website mentioned. It is sometimes possible to find a relocated web page by typing in the address of the home page for a website in the URL window of your browser.

Hachette UK's policy is to use papers that are natural, renewable and recyclable products and made from wood grown in well-managed forests and other controlled sources. The logging and manufacturing processes are expected to conform to the environmental regulations of the country of origin.

Orders: please contact Hachette UK Distribution, Hely Hutchinson Centre, Milton Road, Didcot, Oxfordshire, OX11 7HH. Telephone: +44 (0)1235 827827. Email education@hachette.co.uk. Lines are open from 9 a.m. to 5 p.m., Monday to Friday. You can also order through our website: www.hoddereducation.co.uk

First published in 2018 by
Hodder Education
An Hachette UK Company,
Carmelite House, 50 Victoria Embankment
London EC4Y 0DZ

Impression number   5   4   3
Year    2024

Cover photo © Scanrail – stock.adobe.com

Illustrations by Aptara, Inc

Typeset in BemboStd 11/13 pts by Aptara Inc.

Printed and bound by CPI Group (UK) Ltd, Croydon, CR0 4YY

ISBN 9781510446755

# Get the most from this book

Everyone has to decide his or her own revision strategy, but it is essential to review your work, learn it and test your understanding. These Revision Notes will help you to do that in a planned way, topic by topic. Use this book as the cornerstone of your revision and don't hesitate to write in it — personalise your notes and check your progress by ticking off each section as you revise.

## Tick to track your progress

Use the revision planner on pages iv and v to plan your revision, topic by topic. Tick each box when you have:

- revised and understood a topic
- tested yourself
- practised the exam questions and gone online to check your answers and complete the quick quizzes.

You can also keep track of your revision by ticking off each topic heading in the book. You may find it helpful to add your own notes as you work through each topic.

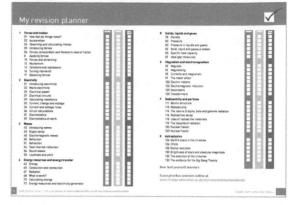

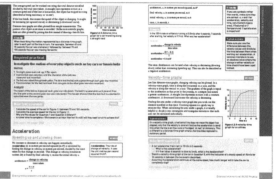

# Features to help you succeed

## Exam tips

Expert tips are given throughout the book to help you polish your exam technique in order to maximise your chances in the exam.

## Typical mistakes

The author identifies the typical mistakes candidates make and explains how you can avoid them.

## Now test yourself

These short, knowledge-based questions provide the first step in testing your learning. Answers are at the back of the book.

## Definitions and key words

Clear, concise definitions of essential key terms are provided where they first appear.

Key words from the specification are highlighted in bold throughout the book.

## Revision activities

These activities will help you to understand each topic in an interactive way.

## Exam practice

Practice exam questions are provided for each topic. Use them to consolidate your revision and practise your exam skills. Answers are online.

## Summaries

The summaries provide a quick-check bullet list for each topic.

## Online

Go online to check your answers to the exam questions and try out the extra quick quizzes at **www.hoddereducation.co.uk/myrevisionnotes downloads**

# My revision planner

## 1 Forces and motion

- 1 How fast do things move?
- 2 Acceleration
- 4 Observing and calculating motion
- 5 Introducing forces
- 6 Forces, acceleration and Newton's laws of motion
- 8 Applying forces
- 10 Forces and stretching
- 12 Momentum
- 14 Collisions and explosions
- 16 Turning moments
- 17 Balancing forces

## 2 Electricity

- 21 Introducing electricity
- 22 Mains electricity
- 25 Electrical power
- 27 Electrical circuits
- 29 Calculating resistance
- 32 Current, charge and voltage
- 34 Current and voltage rules
- 36 Circuit calculations
- 37 Electrostatics
- 38 Electrostatics at work

## 3 Waves

- 42 Introducing waves
- 44 Ripple tanks
- 47 Electromagnetic waves
- 49 Reflection
- 50 Refraction
- 53 Total internal reflection
- 55 Sound waves
- 57 Loudness and pitch

## 4 Energy resources and energy transfer

- 61 Energy
- 63 Conduction and convection
- 65 Radiation
- 66 What is work?
- 68 Calculating energy
- 70 Energy resources and electricity generation

REVISED    TESTED    EXAM READY

**5  Solids, liquids and gases**

REVISED    TESTED    EXAM READY

75  Density

77  Pressure

77  Pressure in liquids and gases

79  Solid, liquid and gaseous states

82  Specific heat capacity

83  Ideal gas molecules

**6  Magnetism and electromagnetism**

88  Magnets

90  Magnetising

91  Currents and magnetism

93  The motor effect

96  Electric motors

98  Electromagnetic induction

99  Generators

101  Transformers

**7  Radioactivity and particles**

106  Atomic structure

108  Radioactivity

109  The nature of alpha, beta and gamma radiation

111  Radioactive decay

113  Uses of radioactive materials

114  The hazards of radiation

115  Nuclear fission

118  Nuclear fusion

**8  Astrophysics**

121  Earth's place in the Universe

121  Orbits

123  Stellar evolution

125  Brightness of stars and absolute magnitude

127  The evolution of the Universe

128  The evidence for the Big Bang Theory

Now test yourself answers

Exam practice answers and quick quizzes at
www.hoddereducation.co.uk/myrevisionnotesdownloads

# Countdown to my exams

## 6–8 weeks to go

- Start by looking at the specification — make sure you know exactly what material you need to revise and the style of the examination. Use the revision planner on pages iv and v to familiarise yourself with the topics.
- Organise your notes, making sure you have covered everything on the specification. The revision planner will help you to group your notes into topics.
- Work out a realistic revision plan that will allow you time for relaxation. Set aside days and times for all the subjects that you need to study, and stick to your timetable.
- Set yourself sensible targets. Break your revision down into focused sessions of around 40 minutes, divided by breaks. These Revision Notes organise the basic facts into short, memorable sections to make revising easier.

REVISED ☐

## 2–6 weeks to go

- Read through the relevant sections of this book and refer to the exam tips, summaries, typical mistakes and key terms. Tick off the topics as you feel confident about them. Highlight those topics you find difficult and look at them again in detail.
- Test your understanding of each topic by working through the 'Now test yourself' questions in the book. Look up the answers at the back of the book.
- Make a note of any problem areas as you revise, and ask your teacher to go over these in class.
- Look at past papers. They are one of the best ways to revise and practise your exam skills. Write or prepare planned answers to the exam practice questions provided in this book. Check your answers online and try out the extra quick quizzes at **www.hoddereducation.co.uk/ myrevisionnotesdownloads**
- Use the revision activities to try out different revision methods. For example, you can make notes using mind maps, spider diagrams or flash cards.
- Track your progress using the revision planner and give yourself a reward when you have achieved your target.

REVISED ☐

## One week to go

- Try to fit in at least one more timed practice of an entire past paper and seek feedback from your teacher, comparing your work closely with the mark scheme.
- Check the revision planner to make sure you haven't missed out any topics. Brush up on any areas of difficulty by talking them over with a friend or getting help from your teacher.
- Attend any revision classes put on by your teacher. Remember, he or she is an expert at preparing people for examinations.

REVISED ☐

## The day before the examination

- Flick through these Revision Notes for useful reminders, for example the exam tips, exam summaries, typical mistakes and key terms.
- Check the time and place of your examination.
- Make sure you have everything you need — extra pens and pencils, tissues, a watch, bottled water, sweets.
- Allow some time to relax and have an early night to ensure you are fresh and alert for the examinations.

REVISED ☐

## My exams

**Edexcel International GCSE (9–1) Physics Paper 1**

Date:..................................................................

Time:..................................................................

Location: ............................................................

**Edexcel International GCSE (9–1) Physics Paper 2**

Date:..................................................................

Time:..................................................................

Location: ............................................................

# 1 Forces and motion

## How fast do things move?

### Average speed

Two measurements must be made to find the average speed: the total distance travelled must be divided by the time taken. In science, we tend to use metres (m) for distance and seconds (s) for time: this gives a speed in metres per second (m/s). In everyday life, kilometres per hour (km/h) and miles per hour (mph) are often used instead.

$$\text{average speed} = \frac{\text{total distance travelled}}{\text{time taken}}$$

$$v = \frac{d}{t}$$

> speed, $v$, in metres per second, m/s
>
> distance, $d$, in metres, m
>
> time, $t$, in seconds, s

**Example**

A runner moves a distance of 500 metres in 200 seconds. Find the average speed in m/s.

Answer

$$\text{average speed} = \frac{\text{total distance travelled}}{\text{time taken}}$$

$$\text{average speed} = \frac{500}{200}$$

$$\text{average speed} = 2.5\,\text{m/s}$$

**Exam tip**

Converting values you are given into standard units, such as metres and seconds, will stop you making mistakes when doing calculations in your exam.

The runner might have been moving faster at some points and slower at others. This is why the answer is an **average** speed. Even short journeys involve starting and stopping; your journey to school each morning is an everyday example.

### Speed and velocity

Scientists make a distinction between **speed** (which can be motion in any direction or combination of directions) and **velocity** (which is motion in a particular direction). In everyday life the two are often mixed up, but direction matters in scientific measurements; 20 m/s is a speed, but 20 m/s due North is a velocity.

> **Speed**: Distance travelled per unit of time
>
> **Velocity**: Speed in a defined direction

The same symbol (usually $v$) is used for both speed and velocity and the value may not change; it is the presence or absence of a direction that tells you which is meant.

### Distance–time graphs

If an object or person is moving in a straight line, a distance–time graph can be drawn. Time is along the horizontal or $x$-axis and distance is along the vertical or $y$-axis. The gradient of the graph is equal to the speed at that point in the journey, so a steeper line means a greater speed.

The average speed can be worked out using the total distance travelled divided by the total time taken. A straight line represents motion at a constant speed and if the line is horizontal, the object has stopped moving (described as being stationary or 'at rest').

If the line bends, this means the speed of the object is changing. It might be increasing (an upward curve) or decreasing (a downward curve).

Distance–time graphs are often plotted from simple measurements. The position of an object at set times is recorded. Unlike most scientific graphs, these are often plotted by joining the dots instead of drawing a best-fit line.

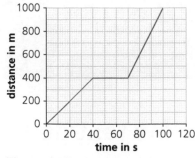

**Figure 1.1 A distance–time graph for a car travelling along a straight road**

### Exam tip

When describing the motion represented by a distance–time graph, refer to each part of the line in turn. For example, 'between 40 and 70 seconds the car was stationary' followed by 'between 70 and 100 seconds the car was moving forwards'.

## Required practical

### Investigate the motion of everyday objects such as toy cars or tennis balls

#### Method

1 Two light gates were set up, 0.5 m apart.
2 A tennis ball was selected, and the diameter of the ball was measured and recorded.
3 The ball was rolled through both gates. The time that the ball took to pass through each gate was recorded.
4 The time taken for the ball to travel from one gate to the other gate was also recorded.

#### Analysis

The speed of the ball as it passed each gate was calculated. The ball's acceleration as it passed from the first gate to the second gate was also calculated. The results showed that the ball had accelerated in speed between the two gates.

## Now test yourself

TESTED

1 Calculate the speed of the car in Figure 1.1 between 70 and 100 seconds.
2 Calculate the average speed for the car in Figure 1.1.
3 Why are the values for Question 1 and Question 2 different?
4 A runner aims to complete a 10 km event in an hour. How fast (in m/s) will they need to run to achieve this?

Answers on page 131

# Acceleration

## Speeding up and slowing down

REVISED

No increase or decrease in velocity can happen immediately. **Acceleration** ($a$) in metres per second squared (m/s$^2$) is calculated by finding the change in velocity in metres per second, divided by the time taken for that change in seconds. The change in velocity (sometimes written $\Delta v$) is found by final velocity $v$, minus the initial velocity $u$.

**Acceleration:** The rate of change of velocity. It uses the unit metres per second squared (m/s$^2$).

$$\text{acceleration} = \frac{\text{change in velocity}}{\text{time taken}}$$

$$a = \frac{v - u}{t}$$

acceleration, *a*, in metres per second squared, m/s²

final velocity, *v*, in metres per second, m/s

initial velocity, *u*, in metres per second, m/s

time, *t*, in seconds, s

**Example**

In the 100m race an athlete is running at 8.5 m/s after 4 seconds; 7 seconds after starting, her velocity is 9.7 m/s. What was her acceleration?

Answer

$$acceleration = \frac{change\ in\ velocity}{time\ taken}$$

$$acceleration = \frac{(9.7 - 8.5)}{3}$$

$$acceleration = 0.4\ m/s^2$$

The term deceleration can be used when velocity is decreasing (slowing down) rather than increasing (speeding up). This can also be described as a negative acceleration.

## Velocity–time graphs

REVISED

Just like distance–time graphs, changing velocity can be plotted. In a velocity–time graph, time is along the horizontal or *x*-axis, and the velocity is along the vertical or *y*-axis. The gradient of the graph is equal to the acceleration at that point in the journey, so a steeper line means a greater acceleration. A straight line represents motion with a constant acceleration. A downward line means the velocity is decreasing.

Finding the area under a velocity–time graph lets you work out the distance travelled in that time. Counting squares is a quick way to estimate this. When calculating the area under a graph, it is usually helpful to divide it into rectangular and triangular sections so that each area can be calculated individually.

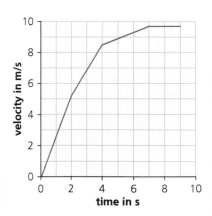

**Figure 1.2 A velocity–time graph for an athlete**

## Now test yourself

TESTED

5 A car accelerates from rest to 15 m/s in 3 seconds.
  (a) What is the acceleration?
  (b) If it then takes 6 seconds to slow to 3 m/s, what is the deceleration?
6 Sketch a velocity–time graph of the car in Question 5, with the inclusion of a steady speed of 15 m/s for 10 seconds in between the two events described.
7 Assuming the deceleration continues at the same speed, how much longer will it take the car to completely stop?

Answers on page 131

# Observing and calculating motion

## Light gates and ticker timers

REVISED

No matter what equipment is used to investigate speed (including frames of a video recorded with a camera), the important equation to consider is the same:

$$\text{speed} = \frac{\text{distance}}{\text{time}}$$

For a light gate, the time is measured for the object with a known length to pass through. For a ticker timer, the time change between each mark is the same, for example 0.02 s, and the distance can be measured with a ruler.

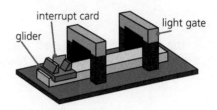

**Figure 1.3 Light gate**

> **Exam tip**
>
> For any measurements or diagrams of a moving object, it should be possible to work out values for distance and time, so speed and/or acceleration can be calculated. Look for clues and add your thoughts, either to the diagram or as bullet points in your notes. These notes can gain partial credit even if you do not get the final answer right.

## Equations of motion

REVISED

An important equation links initial and final speed, acceleration and distance travelled. This is a useful equation because it does not include any value for the time taken, which is often hard to measure. If three values are known, the fourth value can be worked out. If an object starts or finishes 'at rest' (stationary) then the relationship becomes much simpler.

$$v^2 = u^2 + (2 \times a \times s)$$

final velocity, $v$, in metres per second, m/s

initial velocity, $u$, in metres per second, m/s

acceleration, $a$, in metres per second squared, m/s$^2$

distance travelled, $s$, in metres, m

> **Example**
>
> A car accelerates at 5 m/s$^2$ for a distance of 30 m. What is the final velocity if the initial velocity was 3 m/s?
>
> Answer
>
> $v^2 = u^2 + (2 \times a \times s)$
>
> $v^2 = 3^2 + (2 \times 5 \times 30)$
>
> $v^2 = 9 + 300$
>
> $v^2 = 309$
>
> $v = \sqrt{309}$
>
> $v = 17.6\,\text{m/s}$ (to 1 d.p.)

> **Typical mistake**
>
> Note that $s$, the symbol for distance, is easily confused with s, the abbreviation for the SI unit of time (seconds). Some students accidentally use it to represent speed rather than the accepted abbreviation $v$. Make sure you are using and interpreting your symbols properly.

> **Revision activity**
>
> Describe your movement between two points using correct terms. Any straight-line journey could be shown using a distance–time or velocity–time sketch. Describe what is happening at different points on the graph. Try sketching a motion graph and then describe it to a partner; can they sketch a similar graph from your description, without seeing the original? All of these activities will improve your fluency in describing motion in words, which makes it much easier to choose the correct values for calculations.

Exam practice answers and quick quizzes at **www.hoddereducation.co.uk/myrevisionnotesdownloads**

Answers on page 131

## Now test yourself

TESTED

8    Rearrange or simplify the formula $v^2 = u^2 + (2 \times a \times s)$
     (a)   to find the distance travelled given the other values
     (b)   if the initial velocity is zero.
9    Calculate the final velocity of a cyclist who accelerates from rest at $2\,\text{m/s}^2$ along a 100 m track.
10   A trolley of length 10 cm rolls through two light gates on a flat table. The first reading is 0.1 s and the second is 0.2 s.
     (a)   Work out the speed at each point.
     (b)   Why is it not possible to work out the acceleration of the trolley?

# Introducing forces

## Definitions and examples

REVISED

**Forces** have three possible effects on an object:
- changing its shape (temporarily or permanently)
- changing its speed (acceleration)
- changing its direction of motion.

> **Force**: A push or a pull

We divide forces into those which require contact and those that don't.
- Contact forces involve particles that push against or collide with each other. Tension is a force acting through a stretched object. Friction resists movement between surfaces that are touching.
- Non-contact forces such as gravitational, electrical and magnetic forces are associated with fields (see page 89 for more information). Gravitational forces can only be attractive but magnetic and electrical forces can be attractive (pulling together) or repulsive (pushing apart). The attraction towards the Earth is called weight and is the gravitational force.

The SI unit of force is the newton (N). Large forces can be measured in kilonewtons (kN) instead. If something cannot be measured in newtons, it is not a force.

## Vectors and scalars

REVISED

Force is a **vector quantity** because it has a specific direction, while a **scalar quantity** has only size or magnitude, but no direction. The direction in a vector quantity might be expressed as left or right, forwards or backwards, or by using a compass direction such as 'due East'.

> **Vector quantity**: A quantity like velocity that has a specified direction as well as size or magnitude
>
> **Scalar quantity**: A quantity like speed that has size or magnitude

## Drawing, adding and subtracting forces

REVISED

Arrows are drawn to show forces. The length of the arrow represents the size or magnitude, so a longer arrow means a larger force. In most diagrams the arrow will start where the force is caused, for example this will be at the centre of an object for weight. If one object pushes against another the force should be shown where they meet. An object which has a motor, or something similar which causes motion, may have a force labelled 'thrust' or 'driving force'.

friction, 400 N      thrust, 400 N

**Figure 1.4 Thrust**

If forces are in exactly the same direction, they can be added together. If they are in exactly opposite directions, it is the difference between them that is important. If they are of equal magnitude but in opposite directions, we say they are balanced.

If an overall or **resultant force** is acting on an object, it will accelerate in the direction of that force. If a resultant force acts on a moving object, it will speed up or slow down depending on the direction of the force.

The opposite is also true. In other words, if an object is not accelerating, there cannot be a resultant force acting on it. For example, the weight of a table acts on it down towards the floor. The table does not accelerate, so there must be an equal force acting up from the floor. This is called the normal contact force.

If two or more forces act on an object, but are balanced, then there may be a change of shape even though there is no acceleration. See page 7 for more detail.

> **Resultant force:** The combination of two or more forces in a line. Like other forces, it will have both magnitude and direction. It may be zero.

## Now test yourself

TESTED

11  Add the missing description and value for the forces in the diagram, given that the resultant force is zero.

700N

thrust

200N, friction

Figure 1.5

12  A rubber ball is at rest until pushed so it rolls along the floor and bounces off the wall. Explain the three possible effects of forces that can be observed.

Answers on page 131

# Forces, acceleration and Newton's laws of motion

## Balanced forces

REVISED

If the forces are of equal magnitude and in opposite directions, they are said to be balanced. This means the resultant force is zero.

Forces can be balanced for a stationary or moving object. For stationary objects, if the force is balanced, it stays stationary. For moving objects, if the force is balanced, it will keep moving in the same direction at the same speed.

Newton's first law is often stated as 'When the resultant force on an object is zero, the forces are balanced and the object does not accelerate.'

Remember that forces like friction oppose motion and, therefore, may balance the force pushing an object forwards. The weight of an object may be balanced by tension (if suspended) or by the normal contact force from the floor, sometimes called the reaction force.

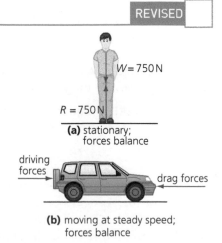

$W = 750\,N$

$R = 750\,N$

**(a)** stationary; forces balance

driving forces

drag forces

**(b)** moving at steady speed; forces balance

**Figure 1.6 Balanced forces**

# Unbalanced forces

If forces are not balanced, there will be acceleration in the same direction as the resultant force. If the object is stationary, it will start to move. If it is already moving, the effect will depend on whether the resultant force is in the same direction as the existing motion or if it is in the opposing direction.

Friction and drag on a moving object, if unbalanced by thrust, will make it decelerate and eventually stop.

# Force, mass and acceleration

The acceleration $a$ depends on two variables: the magnitude of the resultant force $F$ and the mass $m$ of the object. Newton's second law states that 'Acceleration is proportional to the resultant force.'

resultant force = mass × acceleration

$$F = m \times a$$

> resultant force, $F$, in newtons, N
>
> mass, $m$, in kilograms, kg
>
> acceleration, $a$, in metres per second squared, $m/s^2$

## Example

A car of mass 1500 kg must accelerate from rest at $2\,m/s^2$. What resultant force must be exerted?

Answer

$F = m \times a$

$F = 1500 \times 2$

$F = 3000\,N$ or $3\,kN$

## Now test yourself

13 Work out the resultant force in each case and state the effect (if any) on the motion.

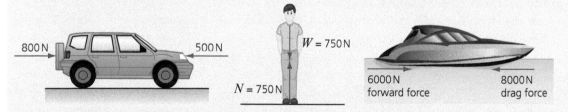

Figure 1.7 **What is the overall force in each case?**

14 Use Newton's laws to explain why a moving car slows down if the engine cuts out, even if the brakes are not used.

15 An engine exerts a forwards force of 40 kN on a train of mass 200 000 kg. What is the acceleration?

Answers on page 131

# Applying forces

## Weight and mass

Weight is a force that acts on objects with mass. It acts towards the Earth because of gravity and it is proportional to the mass. The strength of the Earth's gravitational field, g, is 10 N/kg. This means that a force of 10 N acts on each kilogram.

weight = mass × gravitational field strength
$$W = m \times g$$

weight, $W$, in newtons, N

mass, $m$, in kilograms, kg

gravitational field strength, $g$, is 10 newtons per kilogram, N/kg (on Earth)

### Example

A student with a mass of 45 kg is on Earth. What is their weight?

Answer

$W = m \times g$

$W = 45 \times 10$

$W = 450\,\text{N}$

In other places in the solar system, g will have a different value. On Mars it is approximately 4 N/kg, so an object will have a smaller weight there than on Earth even though the mass is the same.

## Falling and parachuting

If there is no **air resistance**, all objects fall at the same rate. Near the Earth, a falling object with no air resistance would accelerate at 10 m/s².

An object accelerates as it falls until the force of air resistance opposing its motion is equal to the weight. Objects that have a small weight or a large area are most affected.

A skydiver has a constant mass, so their weight does not change. Speed and air resistance increase as they fall, until the forces pushing up and pushing down balance, and the velocity becomes constant. If the parachute opens, the air resistance is suddenly much higher, so the skydiver will decelerate until the forces balance again.

**Air resistance**: The force that opposes motion for an object travelling through a gas. It is larger for objects that have a large area or a high speed.

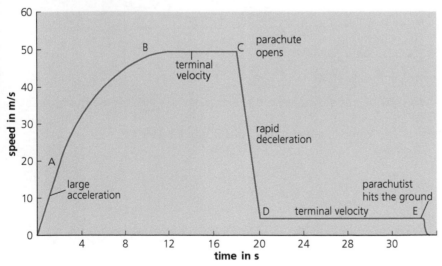

Figure 1.8 Speed–time graph for a skydiver

The skydiver has balanced forces for two parts of the graph; once when travelling quickly with a small surface area, and the second time when they had a larger surface area and a lower speed. Falling at a constant speed as a result of balanced forces is called **terminal velocity**.

> **Terminal velocity**: The speed for a falling object at which the force down (weight) and force up (air resistance or drag) are balanced. The actual value depends on the weight and area of the object, and the properties of the air.

## Driving safely

### Stopping distances

For a moving vehicle like a car, it takes longer to stop (in time and distance) at higher speeds. The total distance travelled while slowing from a certain speed is called the stopping distance and there are two parts you need to consider:

**stopping distance = thinking distance + braking distance**

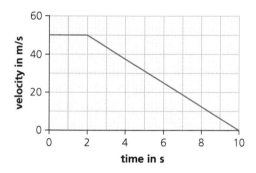

Figure 1.9 Velocity-time graph for a car that is stopping

Anything that increases the driver's reaction time (distractions such as mobile phones, tiredness and the effect of alcohol or other drugs) will increase the thinking distance, so the overall stopping distance will be greater. Higher speed also means a greater thinking distance, because the vehicle travels further in the same time.

thinking distance = speed × reaction time

> **Stopping distance**: The total distance travelled while slowing from a certain speed.
>
> **Thinking distance**: The distance travelled by the car while the driver is reacting to the hazard.
>
> **Braking distance**: The distance travelled between the time the car starts to decelerate and the time it comes to a complete stop.

**Example**

An alert driver travelling at 10 m/s reacts in 0.6 seconds. How far does the car travel before braking starts?

Answer

thinking distance = speed × reaction time

thinking distance = 10 × 0.6

thinking distance = 6 m

**Exam tip**

Most questions on stopping distance will ask you to analyse provided data or suggest the effects of possible factors. Be sure you state clearly what is increased or decreased and why. For example, 'wet roads increase the braking distance because there is less friction' is a better answer than 'wet roads make braking distance longer'.

The car cannot stop instantly and if it decelerates less effectively the braking distance will be more, so the overall stopping distance will be greater. Most factors increasing the braking distance are to do with the car, such as having a heavy load, poorly maintained brakes, or less friction between the tyres and the road because of ice, water, mud or worn tyre treads. Higher speeds also mean greater braking distances.

## Now test yourself

16 A Martian rover design has a mass of 900 kg on Earth. Mars has a gravitational field strength of 4 N/kg. What is the:
   (a) weight on Earth
   (b) mass on Mars
   (c) weight on Mars?

17 Which has the lowest terminal velocity; a skydiver falling at a flat angle, one falling headfirst or one with an open parachute? What difference does this make to the skydiver?

18 A driver is travelling at 20 m/s and is distracted by the sound of a notification on their phone, so their reaction time is 0.9 s. How far does their car travel before they start to brake in an emergency?

19 What effect does ice on the road have on stopping distance and why?

Answers on page 132

# Forces and stretching

## Elastic deformation

If more than one force is applied to an object then the shape can be changed. If there are two forces in opposite directions, the object will be either stretched or compressed depending on whether they are directed away from or towards each other.

If there are three balanced forces on an object, it can be made to bend at the point where the middle force is applied.

Changes in shape can either be temporary (called **elastic deformation**) or permanent (called **inelastic** or **plastic deformation**).

Many objects will be deformed elastically by small forces but become permanently deformed (and damaged) by large forces. A table tennis ball is a good example of this.

> **Elastic deformation**: A temporary change in shape; when the force is removed the object will return to the original size and shape.
>
> **Inelastic** or **plastic deformation**: A permanent change in shape

### Typical mistake

In physics, elastic describes the behaviour of a material, not the material it is made of. (Although it is true that a rubber band shows elastic behaviour under some conditions!)

## Stretching a helical spring

Although almost everything can be stretched or compressed by forces, springs are useful for experiments because they are easy to measure. The stiffer an object is, the harder it is to stretch (or compress). This is described by the **spring constant**, $k$. A spring with a spring constant of 70 N/m will need a force of 70 N to increase the length by 1 m.

> **Spring constant**: Describes how large a force in newtons is needed to increase the length of an object by one metre. It describes the specific object, not the material.

### Exam tip

Make sure that the extension, not length, is used in the equation, and that values are measured in metres, m. A spring that has stretched from 8 to 14 cm has an extension of 0.06 m. It would be wrong to say its extension is 14 cm (which is the length), or that the extension is 6 cm (which is the wrong unit to use).

force = spring constant × extension

$$F = k \times e$$

force, $F$, measured in newtons, N

spring constant, $k$, measured in newtons per metre, N/m

extension, $e$, measured in metres, m

(sometimes $\Delta x$ rather than $e$ is used for extension)

### Example

A spring stretches by 24 cm when a force of 6 N is applied. What is the spring constant, $k$, for this spring?

Answer

$$k = \frac{F}{e}$$

$$k = \frac{6}{0.24}$$

$$k = 25 \text{ N/m}$$

### Revision activity

From memory, describe briefly how you could investigate the relationship between force and spring extension. For each variable in your method, specify the device used for making measurements and include the units. What hazards were identified and how were these controlled? Compare the method you remembered with the one provided in your lessons and add any missing details. Repeat in a week to improve your recall and clarity.

## Limit of proportionality

REVISED

For many materials including springs, the extension is doubled every time the force applied is doubled, until the spring starts to break. This is a directly proportional relationship and is described by **Hooke's law**.

To find out if a material obeys the law, increase the force acting on an object, for example a spring, and measure the extension. The results can then be plotted on a graph. Hooke's law applies until the line on the graph is no longer straight.

Sometimes the force is plotted on the vertical axis and extension along the horizontal axis. Although this is not the normal approach for choosing axes, it means that the spring constant can be calculated by finding the gradient for the straight-line section of the graph.

The limit of proportionality is where the line stops being straight and the increase in length is no longer proportional to applied force. It is sometimes called the elastic limit.

**Hooke's law:** This states that extension of a spring will be proportional to the load up to a maximum force.

## Required practical

### Investigate how extension varies with applied force for helical springs

#### Method

1 A retort stand was set up and a ruler was clamped vertically in place.
2 At the top of the stand a helical spring was attached.
3 The position of the bottom of the spring was measured on the metre rule to give $l_0$.
4 A weight of 1 N was placed on the bottom of the spring.
5 The new position of the bottom of the spring was measured on the ruler to give $l_1$.
6 The extension of the spring was calculated by working out: $l_1 - l_0$.

**7** The experiment was continued by adding further weights and measuring each new position of the bottom of the spring.

### Results

The measurements were recorded and plotted on a graph. The graph showed that for each 1 N weight added, the spring extended a similar amount. At a certain point, the limit of proportionality was reached, and the spring's extension increased beyond the expected proportion.

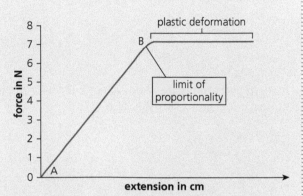

**Figure 1.10 The spring obeys Hooke's law until point B.**

Metal wires show similar behaviour to springs, but the value for the maximum force before the limit is reached will be different in each case. Because any stretching after the limit of proportionality is no longer elastic, wires and springs may not return to their original length when the force is removed. This is inelastic or plastic deformation.

Rubber bands show a different pattern and become stiffer as they get near maximum extension. This is because of the changing alignment of the molecules in the rubber.

## Now test yourself

TESTED

20 A 5 cm spring extends to 7 cm with a force of 4 N. What will the extension be for a force of 8 N, assuming the limit of proportionality has not been reached?

21 (a) Find the spring constant $k$ for the spring in Question 20.
   (b) What is the predicted extension for a force of 1000 N?
   (c) Explain why this may not occur in practice.

22 Explain the difference between elastic and plastic deformation.

Answers on page 132

# Momentum

## Forces and change of momentum

REVISED

A property that helps us understand the movement of an object is momentum. The more momentum something has, the harder it will be to stop it moving.

momentum = mass × velocity

$$p = m \times v$$

> momentum, $p$, in kilogram metres per second, kg m/s
>
> mass, $m$, in kilograms, kg
>
> velocity, $v$, in metres per second, m/s

### Example

A hockey player, mass 55 kg, is running forwards at 5 m/s. What is her momentum?

Answer

$p = m \times v$

$p = 55 \times 5$

$p = 275$ kg m/s forwards

### Exam tip

A stationary object has a momentum of zero, no matter what mass it has.

A single value for momentum seems abstract. Being able to compare the momentum of two objects means we can predict how they affect each other. Momentum has a direction, so like velocity it is a vector quantity. Often one direction is described as positive, and the other direction is said to be negative to make comparisons easier.

When a resultant force acts on an object, it causes acceleration. This means that there is a change of momentum too. The definition of momentum can be combined with Newton's second law (see page 7 for a reminder) to link force with a change of momentum.

$$\text{force} = \frac{\text{change of momentum}}{\text{time}}$$

$$F = \frac{mv - mu}{t}$$

also written as

$$F = \frac{\Delta mv}{t}$$

If the object starts or stops moving, all of the momentum is either gained or lost. If it is a moving object that is speeding up or slowing down, it is important to use the relative change ($\Delta mv$) not the absolute value at the start ($mu$) or end ($mv$) in the calculation.

If an object is slowing down or speeding up, this form of the equation is used to find the force the object experiences. At other times there is a known force applied to an object for a specific amount of time; this means the change in momentum can be calculated.

force, $F$, in newtons, N

change of momentum, $\Delta mv = mv - mu$, in kilogram metres per second, kg m/s

time, $t$, in seconds, s

**Example**

The hockey player above comes to a sudden stop. It takes half a second for this to happen. What force acts on her?

Answer

$$F = \frac{\Delta mv}{t}$$

$$F = \frac{275}{0.5}$$

$$F = 550\,\text{N}$$

## Momentum and safety

REVISED

A moving object that stops, whether gradually or suddenly, will have the same change of momentum. However, the maximum force applied will be greater if the same momentum change occurs in less time. To reduce damage and injury, the change in momentum can be spread out over a longer period of time, which means a smaller deceleration. This, in turn, means that although the force acts for longer, it has a smaller magnitude. Passengers in a car that brakes gradually experience a smaller force and less harm than those in a sudden stop or crash, so the force will potentially hurt less for anyone affected by the force.

Anything that makes a change in momentum less sudden will reduce the maximum force which acts. This can be achieved by part of the object bending or sliding to reduce the velocity more gradually. It can also help to use a material which crushes or compresses gradually, so the time is increased.

**Typical mistake**

Marks are often lost by saying that soft materials are used for slowing down because the 'impact is spread out more'. However, a better explanation is that 'the change of momentum takes place over a longer time, so the force is less'.

## Driving safely 2 (crumple zones and seat belts)

REVISED

Cars provide several examples where the time taken for the change in momentum is made greater, so the deceleration and force both decrease. **Crumple zones** and seat belts mean less damage to the passengers if a collision happens.

**Crumple zones:** The parts of a car designed to buckle, which takes extra time compared to a rigid frame on impact. While this is happening, the areas where the people are sitting are decelerating more slowly and so the force on them is less.

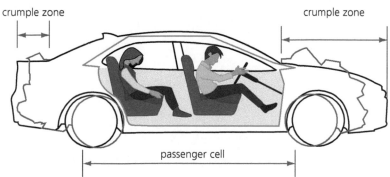

crumple zone      crumple zone

passenger cell

**Figure 1.11 Crumple zones and seat belts in a car increase the time for momentum changes.**

Seat belts are designed to let the wearer move a little during an impact, but slower than if unsecured. Again, this means the deceleration is reduced and the force is less, so damage is less likely. There was a clear drop in injury rates when seat belts were made compulsory in the UK in 1983.

## Now test yourself

23 What is the momentum of a tennis ball of mass 60 g, moving at 20 m/s?
24 Is momentum a vector or scalar quantity? What does this mean?
25 (a) An athlete of mass 65 kg accelerates from 4 to 7 m/s. What is their increase in momentum?
   (b) Why is it not possible to calculate the force needed for this increase in velocity?
26 In a crash at 25 m/s, crumple zones and seat belts in a test car increase the time taken for the driver to come to a stop to 0.4 seconds. Calculate the force acting on a 80 kg driver.

Answers on page 132

# Collisions and explosions

## Conservation of momentum

For a single object, momentum can be increased or decreased when a force acts. In collisions or explosions, the total momentum is conserved unless a force acts to plastically deform one or more of the objects. This means that, unless told otherwise, momentum before a collision is equal to momentum after, as long as all of the objects involved are considered.

   momentum before = momentum after (unless an external force is acting)

**Exam tip**

Describing these situations as 'balanced' rather than 'cancelling out' will show that you understand how vector quantities work.

### Example

At the funfair, two bumper cars, each of mass 400 kg, collide as shown in the diagram below. What is the new velocity of the second car?

Figure 1.12 **Bumper cars before and after colliding**

#### Answer

momentum before $= m_1 v_1 + m_2 v_2$

momentum before $= 400 \times 1.5 + 400 \times 1$

momentum before $= 1000\,\text{kg m/s}$

momentum after must be the same so:

momentum after $= 1000\,\text{kg m/s}$

$m_1 v_1 + m_2 v_2 = 1000$

$0 + m_2 v_2 = 1000$

$v_2 = \dfrac{1000}{m_2}$

$v_2 = \dfrac{1000}{400}$

$v_2 = 2.5\,\text{m/s}$

Because momentum is a vector, it is possible for two colliding objects to have a total momentum of zero. This is similar to the idea that equal and opposite forces on an object give a resultant force of zero.

## Example

Two canoes are moored together in the middle of a calm lake. The canoeists push apart and the first, with a mass of 150 kg, moves at a velocity of 2 m/s towards the North. The second has a velocity of 2.4 m/s towards the South. What is the mass of the second canoeist?

at rest

### Answer

Before, the total momentum was zero. So the total afterwards must be zero too. To understand the equation, one direction must be defined as positive and the other must be defined as negative. Taking North as positive:

$$m_1 v_1 + m_2 v_2 = 0$$

$$(150 \times 2) + (m_2 \times -2.4) = 0$$

$$300 = 2.4 \times m_2$$

$$m_2 = \frac{300}{2.4} = 125\,\text{kg}$$

2 m/s

2.4 m/s

**Figure 1.13 Canoes before and after being pushed apart**

Another common situation involves two objects sticking together or splitting apart. Calculations will still be based around the fact that momentum before is the same as momentum after, but remember to add the masses when treating them as one combined object.

## Newton's third law

REVISED

Although we often only think about one force at a time, in reality there are no single forces. When a student leans against a wall, applying a force on it, then there must be an equal and opposite force acting on the student. This is Newton's third law.

## Now test yourself

TESTED

27 A school lab trolley is loaded to have a mass of 2 kg, and is at rest. A second trolley, mass 1 kg, collides with it at a velocity of 0.8 m/s and stops. What is the new velocity of the first trolley?
28 A student, mass 45 kg, runs at 3 m/s, jumps and lands on a stationary skateboard, mass 5 kg. Assuming they don't fall off, what is their new (rolling) velocity?
29 A swimmer exerts a force of 60 N on the water with each stroke. What force acts on the swimmer?

Answers on page 132

### Typical mistake

It is easy to think that because thrust and air resistance for a car moving at constant speed are equal and opposite, this is an example of Newton's third law. However, the forces must be acting on different objects, so a better example would be that the air resistance on a car has an equal and opposite (total) force acting on the particles in the air it collides with.

# Turning moments

When a force causes something to turn rather than accelerate, both the size **and** the position of the force are important. The greater the perpendicular distance from the fixed point or pivot, the larger the effect of that force. This is called a **turning moment**.

turning moment = force applied × perpendicular distance of the line of force from the pivot

**Turning moment:** The effect of a force which is exerted around a point. Turning moments are vector quantities, and the direction is usually described as being clockwise or anti-clockwise. Moments are measured in newton metres (Nm).

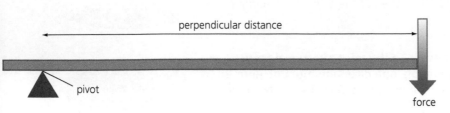

**Figure 1.14 Turning moment**

## Turning forces in action

REVISED

### Example

What turning moment is applied to the door if the force is 80 N?

**Answer**

turning moment = force applied × perpendicular distance

turning moment = 80 × 0.7

turning moment = 56 Nm

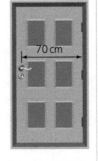

70 cm

Figure 1.15

**Exam tip**

It often helps to draw a diagram for moments questions if there isn't one provided. Label the load and effort, which will be forces in newtons, as well as the distance from each to the pivot.

**Typical mistake**

Don't mix up the units for moments (newton metres, Nm) with those for pressure (newtons per metre squared, N/m²) or spring constant (newtons per metre, N/m).

**Exam tip**

Despite the similarity of the words, moments and momentum are not linked. Make sure you use the right word to get the mark.

## Lifting loads

REVISED

Whenever a load is lifted, there is a turning moment on the object doing the lifting. This is why it is much harder to lift an object at arms' reach than it is when it's closer to your body. This distance is the **working radius**.

If a load is a long way from the object lifting it, the large moment may cause the object to tip over.

**Working radius:** The perpendicular distance from the load of a crane (which exerts a force equal to the weight) to the base of the arm, which is the pivot. As the arm extends, the turning effect becomes greater (because the perpendicular distance has increased). There will be a maximum safe load for each possible length of arm.

## Centre of gravity

The mass of most objects is spread throughout their volume, but they can be treated as if all the mass is at a single point, called the **centre of gravity**. This means all the weight is treated as a single force in one place.

Objects that are uniform, like a football or a steel cube, have a centre of gravity near the centre of the object. If the object is made of a mixture of materials, or is an irregular shape, the position of the centre of gravity is harder to predict.

An object that is free to move will balance itself eventually because the turning moment of the object's weight will cause rotation, until the centre of gravity is as low as it can be. If one part of the object is held up and it is allowed to turn, the centre of gravity will always be directly beneath this part.

If the centre of gravity of an object is high up, such as in a vehicle or building, the object may tip easily and so it is said to be unstable.

> **Centre of gravity:** The point through which an object's weight acts

> **Revision activity**
>
> List all the quantities mentioned so far in this chapter. For each one, record the symbol, measuring device, the units and their abbreviation. Check your notes to complete the list then test yourself each day until you can remember all the details. Leave it a week, and then repeat.

### Now test yourself

TESTED

30 A force of 300 N on the end of a lever has a turning moment of 750 Nm. What is the perpendicular distance to the pivot?

31 For a working radius of 12 m, a crane has a maximum load of 65 tonnes. A manager suggests that at 24 m it can lift twice the load. Are they correct or incorrect? Explain your reasons.

Answers on page 132

# Balancing forces

## The principle of moments

REVISED

A clockwise turning moment will cause an object to turn clockwise. An anti-clockwise turning moment will cause it to turn anti-clockwise. If the clockwise and anti-clockwise turning moments are the same, then there will be no overall effect and the moments are said to be in **equilibrium**.

A simple example is a playground see-saw; the further a person sits from the pivot, the larger the effect they have. A small child can balance a large adult if they sit at different distances from the pivot. Their weights are in the same plane, parallel rather than opposite, and they have different magnitudes. The turning effects of the two forces are equal in size and opposite in direction.

In this example, the forces, along with the people, will also be balanced. This is an example of Newton's first law. It means that for a see-saw, we can work out the force that is pushing upward from the pivot which is balancing each of the weights.

> **Equilibrium:** The point where the sum of the clockwise moments is equal to the sum of the anti-clockwise moments. The forces also balance.

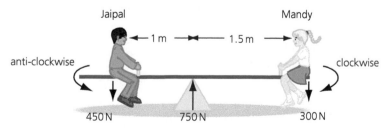

**Figure 1.16 Force at a pivot**

# Moments in action

The idea of equilibrium helps us work out how to move a load. The turning effect of the effort only needs to be slightly bigger than the turning effect of the load (in the opposite direction) to make a difference.

> **Exam tip**
>
> When describing a situation at equilibrium, consider the turning moment of each force in turn, taking care to separate clockwise and anti-clockwise moments.

**Example**

What force at the handles is needed to just lift the wheelbarrow in the diagram?

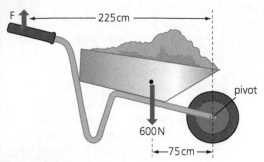

Figure 1.17

Answer

At equilibrium:

clockwise moments = anti-clockwise moments

$$F_1 d_1 = F_2 d_2$$

$$F_1 = \frac{F_2 d_2}{d_1}$$

$$F_1 = \frac{600 \times 0.75}{2.25}$$

$$F_1 = 200\,\text{N}$$

To stop objects tipping over, a counterbalance can be used. A counterbalance is often an extra force that causes a turning moment to oppose the load, so the object is balanced. The magnitude of the force and the distance from the pivot are chosen specifically so the object is at least in equilibrium.

# Forces on beams

The simplest possible model of a bridge or shelf is a light beam (a straight line supported at each end). For any object placed on the beam, it is possible to calculate the turning moment at either end to find how much of the weight is supported by each end.

**Example**

A heavy vase, weight 30 N is placed on a 50 cm shelf that is fixed at each end. If it is 10 cm from one end, what will the force exerted at the other end be?

**Answer**

If $F_2$ is the upward force at the far end to counter $F_1$ the weight of the vase.

$$F_1 d_1 = F_2 d_2$$

$$F_2 = \frac{F_1 d_1}{d_2}$$

$$F_2 = \frac{30 \times 0.1}{0.5}$$

$$F_2 = 6\,\text{N}$$

## Now test yourself

TESTED ☐

32 What is the unit of a turning moment?

33 A crane with an arm of 12 m lifts a weight of 4000 N. What counterbalance must be placed at 2 m on the opposite side of the pivot to create equilibrium?

34 A bridge supports a car of weight 15 kN between two supports. What force is exerted by each end of the bridge because of the car in the centre?

Answers on page 132

## Summary

- average speed = $\dfrac{\text{total distance travelled}}{\text{time taken}}$
- Speed is a scalar quantity (magnitude only) but velocity is a vector (magnitude and direction).
- The gradient of a distance–time graph is equal to the speed.
- Acceleration is the rate of change of velocity:
  acceleration = $\dfrac{\text{change in velocity}}{\text{time taken}}$
- The area under a velocity–time graph is equal to the distance travelled.
- Light gates and ticker timers are used to measure speed; both use measurements of distance travelled and time taken to do this.
- Properties of an object travelling with constant acceleration:
  $v^2 = u^2 + (2 \times a \times s)$
- Forces cause objects to change shape, speed or direction of motion.
- When more than one force acts, the overall or resultant force is often useful.
- Newton's first law states that when the resultant force on an object is zero, the object remains at rest or moves with a constant speed in a straight line (there is no acceleration).
- Newton's second law states that acceleration is proportional to resultant force:
  resultant force = mass × acceleration
- Weight is a force on an object that acts towards the Earth because of gravity:
  weight = mass × gravitational field strength

- Air resistance is a force that resists motion of an object in atmosphere; it increases with both speed and surface area.
- stopping distance = thinking distance + braking distance
- Thinking distance is increased if the driver has a high reaction time, for example because of tiredness, distraction or some drugs including alcohol.
- Braking distance is increased if the car tyres or brakes are poorly maintained, if the road is slippery, for example because of ice or snow, or if the vehicle is overloaded.
- An object that stretches because of a force will obey Hooke's law up to the elastic limit:
  force = spring constant × extension
- momentum = mass × velocity
- The force acting in a collision can be reduced if the time taken can be increased, for example by seat belts or crumple zones:
  force = $\dfrac{\text{change of momentum}}{\text{time taken}}$
- Unless an external force acts, the momentum before and after a collision will be the same.
- Newton's third law states that whenever two objects interact, the forces they exert on each other are equal and opposite.
- turning moment = force applied × perpendicular distance of the line of force from the pivot
- When an object is in equilibrium, the clockwise and anti-clockwise moments are equal.

# Exam practice

1  (a) State the normal unit of speed. [1]
   (b) A car travels 800 m in 15 seconds. Calculate how fast it is travelling. [3]
   (c) Calculate whether this is faster or slower than one that travels 6 km in 2 minutes. [2]
2  (a) If a box is at rest on the floor and has a weight of 120 N, calculate the normal contact force upwards. [1]
   (b) Using a gravitational field strength of 10 N/kg, calculate the mass of the box. [3]
   (c) On the Moon, the gravitational field strength is 1.7 N/kg. What would the weight of the box be if it was placed on the Moon? [2]
3  The figure shows a velocity–time graph for a ball bearing, weight 0.5 N, which is falling from rest.

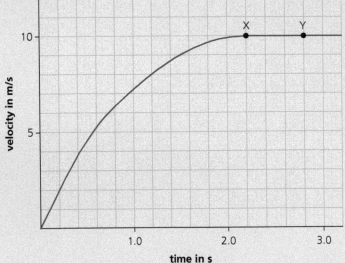

(a) Calculate the average acceleration in the first 1.2 seconds. [3]
(b) Calculate the distance travelled between the times marked X and Y on the graph. [3]
(c) State the magnitude of the air resistance acting on the ball bearing after 2.2 seconds. [1]
(d) Explain the shape of the graph. [4]

4  A children's playground is designed with foam tiles underneath the apparatus. Explain how these will reduce the chance of injury for falling children. [3]
5  A student is asked to investigate the effect of changing force on the extension of a steel spring.
   (a) Identify the independent variable in the investigation shown in the figure on the right. [1]
   (b) Define the limit of proportionality for a spring. [1]
   (c) Until the spring is stretched beyond this point, the line of best fit is straight. Explain how you would use the line to find the value of the spring constant, $k$. [2]
   (d) Not all the points lie exactly on the line of best fit. Explain how you could change the method to improve this. [2]
6  A child, weight 300 N, sits 1.8 m from the pivot of a balanced see-saw.
   (a) Calculate the turning moment of the child. Give the unit in your answer. [3]
   (b) State the principle of moments. [1]
   (c) An adult is sitting 60 cm from the pivot on the opposite side. Calculate their weight given that the adult and child balance. [3]

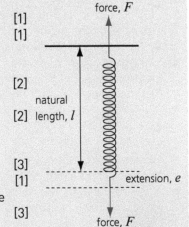

**Answers and quick quizzes online**                                    ONLINE

# 2 Electricity

## Introducing electricity

### Electrical transfer of energy

An electrical circuit transfers energy between stores, for example from the chemical store of a cell to the thermal store of a lamp filament (see page 61 for more information on energy stores). There are different models we can use to explain how electricity works. Observations and measurements made in the school lab can only be explained properly by considering what is happening in the components at a particle level.

> **Exam tip**
>
> It is best to use a specific quantity in answers to questions about electrical circuits. For example, talking about one of the variables with a symbol and a unit rather than just writing 'electricity', which is a description of the whole process: 'An extra cell in a circuit might increase the voltage by 1.5V', instead of 'An extra cell in a circuit might provide more electricity'.

### Voltage and current

In electrical circuits the **electrical cell** is what makes things happen. It provides **voltage** which causes movement of tiny charged particles through the wires. The rate of flow of charge is what we call **current**. Current transfers energy from the chemical store of the cell to other components that might be in a circuit. For example, it would increase the thermal store of a heating element. The hot filament then transfers energy to the thermal store of the room by several processes.

A larger voltage will cause a larger current and so energy is transferred more quickly. In the example of a lamp, this will usually mean the bulb will be brighter.

> **Electrical cell**: A component that contains chemical compounds which react to provide a voltage
>
> **Voltage**, $V$, measured in volts (V): The cause of current in a circuit. Different energy supplies provide different voltages. For example, a new AA cell provides 1.5V and the UK mains supply provides 230V.
>
> **Current**, $I$, measured in amperes or amps (A): The movement of particles in the wire. These moving particles transfer energy around a complete circuit.

> **Exam tip**
>
> If you see the unit mA, remember that 1000mA = 1A.

## Electrical hazards

Electricity is a very useful way of transferring energy, but it can be dangerous too. This is why electrical pylons and railway lines are marked with 'Danger: High Voltage' signs. If an electrical current flows through a person instead of a wire, it can damage the heart or brain. Even getting close to a high-voltage supply is dangerous because a spark might jump from the supply to the person and complete the circuit.

Large currents can also cause a huge temperature increase and may start a fire.

### Now test yourself

1  (a)  A bulb is dimly lit when in a circuit with a single 1.5V AA cell. What voltage would be supplied by three AA cells together?
   (b)  What effect would an increase in current have on a bulb in the circuit?
2  Why should you not get close to high-voltage supplies like pylons?
3  Which is the cause of electricity, voltage or current?
4  The current in a circuit is displayed as 250mA. What is this value in amps?

Answers on page 132

# Mains electricity

## Alternating and direct current

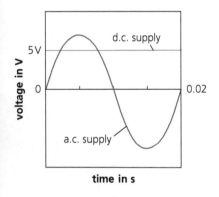

**Figure 2.1 Direct (d.c.) and alternating (a.c.) supply voltages**

If the voltage supplied by a cell or power supply is always in the same direction, so is the current caused by it. This is called a **direct current**. Cells and batteries are a source of direct current. If the voltage changes direction, an **alternating current** is produced.

The graph in Figure 2.1 shows how the voltage supplied changes over time for a.c. but is constant for d.c. Both supply 5V over time, but for a.c. this is an average between high and low values during the cycle.

> **Direct current (d.c.):** When voltage and current are in a constant direction around the loop. It is very important that cells and batteries are correctly orientated in a circuit.
>
> **Alternating current (a.c.):** The voltage and current change direction in a circuit many times each second.

# Mains supply

REVISED

Mains electricity is always a.c. Each complete cycle, with voltage changing direction and back again (as shown in Figure 2.1), takes $\frac{1}{50}$ of a second for the UK mains supply. We say the **frequency** is 50 **hertz** (Hz). The voltage is 230V.

> **Frequency**: The number of complete cycles each second, measured in hertz. This idea is also used to describe waves (see page 43 for more detail).
>
> **Hertz (Hz)**: The unit of frequency, describing the number of complete cycles per second

# Electrical heating

REVISED

Whenever current flows in a material it causes an increase in temperature. Materials with a higher **resistance** heat up more as the charges move through them. To change how a circuit works – for example to change the brightness of a lamp – a resistor with a specific value can be included. A resistor will reduce the current flowing in a circuit. Usually a resistor gets warm as energy is transferred to its thermal store and some are burning hot to the touch. The higher the resistance in the circuit, the harder it is for charges to be pushed through it and so the lower the current.

In the home, many devices involve electrical heating. A resistor heats up when the current flows, and this is used to heat air or water. The element in a kettle or boiler is basically a large value resistor.

> **Resistance, R**: A measure of how hard it is for current to flow through a material or component. The unit is ohms, $\Omega$.

> **Exam tip**
>
> You may see kilo-ohms (k$\Omega$) used in questions. Remember that 1000$\Omega$ = 1 k$\Omega$

# Electricity at home

REVISED

Electricity is very useful in the home or workplace because it can transfer energy to many different components that can convert this energy into all sorts of different forms. However, as electricity is dangerous, several devices may be included in the circuit to make it safer.

## Fuses

Fuses are included in circuits to prevent currents getting dangerously high. Fuses are designed so that the wire inside melts if the heating produced by a high current is too great. When the wire melts, or the fuse 'blows', the circuit is broken so no further current can flow and the electricity transfer will stop. Plugs include a fuse, and they are labelled with the highest current that can flow before breaking or 'blowing'. The fuse is always part of the **live wire** in the circuit or plug, and never the **neutral** one. These two wires make a complete circuit between the power supply and the components.

> **Live wire**: Wire in a circuit with a high voltage compared to the other wires
>
> **Neutral wire**: Wire that completes the circuit; the voltage is always compared to this wire

> **Exam tip**
>
> When answering questions on this topic, remember that the fuse value should always be above the normal current in the circuit. When choosing a fuse, calculate the expected current and then select a fuse that is rated slightly higher.

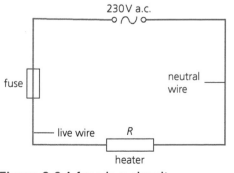

**Figure 2.2 A fuse in a circuit**

## Circuit breakers

Like a fuse, a circuit breaker prevents a high current from flowing. Instead of melting, circuit breakers are designed so that a high current causes an electromagnetic switch to open. This works because of the motor effect (see page 93 for information), with a force acting due to the current. Unlike fuses, circuit breakers are easy to reset once the problem is solved and the current is back to a safe level. Circuit breakers can be set to any value, not just common ones such as a 13 A fuse.

## Earthing and insulation

If a person touches a live wire the current will flow through their body and could hurt or kill them. This can also occur if a device has metal parts on the outside and a live wire is touching the metal part from the inside. To prevent this current from flowing out of the device, an **earth wire** is included in UK sockets, which safely carries the current to the ground. We call this earthing because the connection is to the planet Earth itself.

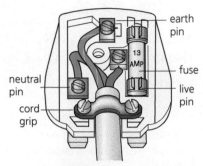

**Figure 2.3 UK plugs have several safety features.**

> **Earth wire:** Wire in a circuit that prevents electrical shock. These are only needed in devices with metal cases.

> **Typical mistake**
>
> It is easy to be vague and say that protective devices and features 'make the current safe'. Make sure you can actually explain the differences between them in detail. Fuses and circuit breakers both limit current, but in different ways. Earthing is used so that if a live wire touches metal parts of a device, the current flows to earth instead of through a person.

Electrical current cannot easily flow through plastic, which is an **insulator**. Devices are often made with a plastic case to prevent current flowing to the outside, which means they don't need an earth connection. This is called **double insulation**, because there are two layers of insulating material.

## Residual-current circuit breaker (RCCB)

In some cases a damaged live wire could touch the outer metal case of an appliance, and a small current would flow to earth without preventing the device from working. Because the current is small, a fuse would not melt. When the device is switched off, the small current could then flow to a person touching the case instead of to the earth. To stop this, a residual-current circuit breaker (RCCB) is sometimes installed. A RCCB detects the current in the live and neutral wires. If the two values are different because there is a small current going to earth, the circuit is disconnected.

> **Insulators:** Materials that do not allow a current to flow, for example plastic or rubber. Materials which do allow a current to flow, like metals, are called conductors. Some materials do not clearly fit into either of these categories.

> **Double insulation:** A protection method which avoids the need for an earth wire. Two layers of insulating material, usually rubber and plastic, are used.

> **Revision activity**
>
> For each variable, practise linking the word, definition, symbol, unit and abbreviation. A good way to do this is to create a flash card for each one. Free mobile apps are also available although index cards are fast and easy to produce.

## Now test yourself

5 List three safety features that can be found in an electrical plug.
6 Would a 60 Hz a.c. supply have greater or fewer numbers of complete voltage cycles compared to a UK mains supply at 50 Hz?
7 Give two advantages for a circuit breaker compared to a fuse.
8 The symbol for current is *I*. What is the unit for current and how is it abbreviated?

Answers on page 132

# Electrical power

## Calculating the power

REVISED ☐

As we know, electrical circuits transfer energy. The amount of energy supplied or transferred per second is called the **power**. For example, a 2.2 kW room heater transfers 2200 J to the thermal store of a room each second.

> **Power, *P*:** The energy supplied or transferred each second. It is measured in joules per second (J/s) or watts (W).

> **Typical mistake**
>
> Students often mix up the ideas of power, current and voltage. Power, current and voltage are connected as both the voltage causing the flow and the current flow itself affect the overall power of the electrical circuit.

You can calculate power in a circuit using the following equation:

power = current × voltage

$$P = I \times V$$

> power, *P*, measured in watts (W)
>
> current, *I*, measured in amps (A)
>
> voltage, *V*, measured in volts (V)

> **Exam tip**
>
> Large power values are often expressed in kilowatts (kW) or megawatts (MW).
> - 1 kW = 1000 W
> - 1 MW = 1000 000 W

> **Typical mistake**
>
> If you confuse symbols when giving an equation as part of an answer, marks may be deducted. For example, $P = C \times V$ would not be accepted as the equation for electrical power. If in doubt, write out the words rather than use incorrect symbols.

> **Example**
>
> What is the power of a mobile phone charger that provides a current of 1.8 A at a voltage of 5 V?
>
> Answer
>
> $$P = I \times V$$
>
> $$P = 1.8 \times 5$$
>
> $$P = 9\,W$$

## Choosing a fuse

REVISED ☐

The current in a device can be found by a rearrangement of the power equation. This value is needed to choose the correct fuse rating which must be above the usual current, but not so high that the current reaches a dangerous level.

$$I = \frac{P}{V}$$

### Example

A kitchen mixer has a power rating of 575 W and is supplied by the UK mains at 230 V. What is the current of this device and should a fuse of rating 3 A or 13 A be used?

Answer

$$I = \frac{P}{V}$$

$$I = \frac{575}{230}$$

$$I = 2.5 \, A \qquad \text{(so a 3 A fuse should be used).}$$

> **Exam tip**
>
> If a question specifies that a device uses the UK mains supply and no other voltage is given, you are expected to remember that this is 230 V.

## Paying for energy

An electricity bill is based on the amount of energy that has been transferred through the devices in a building. To work out how much energy this is, the power rating and the length of time are both needed:

energy = power × time

$$E = P \times t$$

This equation can be combined with $P = I \times V$ for power, giving

energy = (current × voltage) × time

$$E = I \times V \times t$$

> power, $P$, measured in watts (W)
>
> energy, $E$, measured in joules (J)
>
> current, $I$, measured in amps (A)
>
> voltage, $V$, measured in volts (V)
>
> time, $t$, measured in seconds (s)

### Example

A student's circuit, with a power supply of 4.5 V and a current of 0.3 A, is left running for 10 minutes. How much energy is transferred?

Answer

$$E = I \times V \times t$$

$$E = 0.3 \times 4.5 \times 600$$

$$E = 810 \, J$$

### Now test yourself

9  A charger has a voltage of 12 V and a current of 0.5 A.
   (a) What is the power?
   (b) How much energy is transferred in 5 minutes?
10 A circuit transfers 400 J in a minute. What is the power?
11 A toaster has a power of 1.8 kW. What current flows in the toaster?

Answers on pages 132–3

> **Typical mistake**
>
> For this and other electrical equations, it is easy to accidentally use values for time in minutes or hours rather than seconds. One way to avoid this is to get in the habit of converting all values given in a question to standard units.

# Electrical circuits

## Continuous circuit (series)

A **circuit diagram** is drawn with straight wires and symbols to show the position of each component. If there is a single loop, with the current flowing through each component in turn, it is called a **series circuit**. Often the value of the voltage being provided is recorded next to the supply, as in Figure 2.4. The current measured by ammeters will be the same everywhere in the loop.

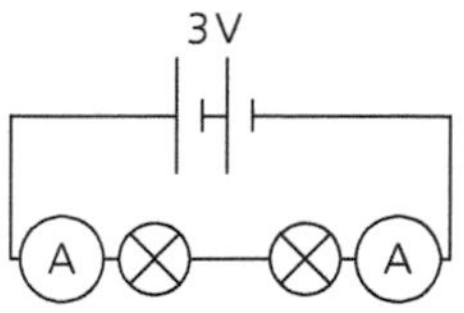

**Figure 2.4 Both ammeters will have the same reading in a series circuit.**

If nothing seems to be happening in a circuit, this is often a sign of a break in the loop. Current cannot flow if the circuit is incomplete, for example because of a broken wire.

> **Circuit diagram**: A simplified layout of the components in a circuit using symbols instead of pictures or words. All wires are drawn as straight lines, so the connections are clear.
>
> **Series circuit**: A circuit with components in one loop and no separate branches

> **Exam tip**
>
> An ammeter measures current (*I*) in amperes or amps (A) and, if included in a circuit, is always part of the loop.

## Controlling the current

If the voltage supplied to a circuit is increased (for example by adding more cells) the current will increase. If the resistance of the circuit is increased (for example by adding more components), the current will be reduced.

Most components work in the same way no matter which way current flows through them. A diode is different and only allows current to flow in one direction, from the positive terminal of the cell to the negative. Some diodes emit light when current flows and are called light-emitting diodes or LEDs. They are useful because they only need a small current to light up.

## Circuit symbols

**Table 2.1 Circuit symbols**

| Description | Symbol |
| --- | --- |
| Conductors crossing with no connection | 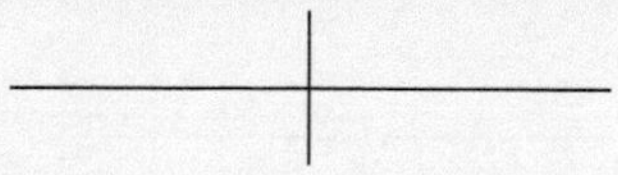 |
| Junction of conductors |  |
| Open switch | 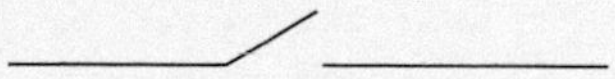 |
| Cell | 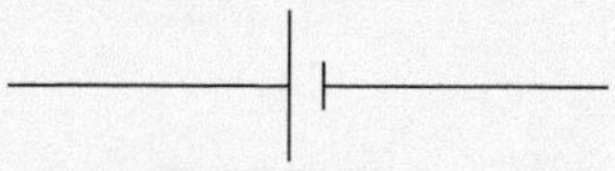 |
| Battery of cells | 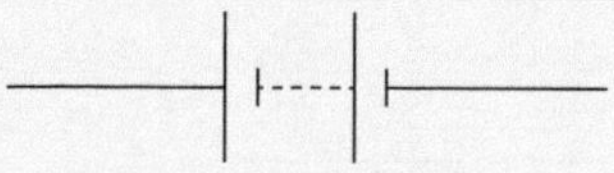 |
| Power supply (d.c.) |  |

| Description | Symbol |
|---|---|
| Power supply (a.c.) | |
| Transformer | |
| Ammeter | |
| Voltmeter | |
| Fixed resistor | |
| Variable resistor | |
| Heater | |
| Thermistor | |
| Light-dependent resistor (LDR) | |
| Diode | |
| Light-emitting diode (LED) | |
| Lamp | |
| Loudspeaker | |
| Microphone | |
| Electric bell | |
| Earth or ground | |
| Motor | |
| Generator | |
| Fuse/circuit breaker | |

**Revision activity**

Learning the symbols in Table 2.1 is a necessary part of preparing for electricity questions in the exams. As well as testing yourself with flashcards, improve your recall by labelling each component whenever you have a practice question that includes a circuit diagram.

12 What component do each of these symbols represent?

(a)   (b)

(c)   (d)

Figure 2.5

13 Will the current increase or decrease in each of the following situations?
   (a) The number of cells is halved.
   (b) A resistor is added to the circuit.
   (c) The number of cells is doubled and the setting on the variable resistor is reduced at the same time.
14 One of the lamps in the circuit shown in Figure 2.4 breaks (see page 27). What happens to the other lamp?

Answers on page 133

# Calculating resistance

## Ammeters and voltmeters   REVISED

Ammeters measure the current passing through the circuit, so they must be part of the loop. They are connected in series.

Voltmeters measure the difference in voltage between two points, so are connected to an existing loop. The two connections are made either side of the component (or combination of components) being tested. This is called a parallel connection or a **parallel circuit**.

> **Parallel circuit:** Has more than one loop. The current in different parts of the circuit will not be the same.

> **Exam tip**
>
> A voltmeter measures voltage (*V*) in volts (V) and is always connected in parallel. If a voltmeter is connected in series the circuit will stop working.

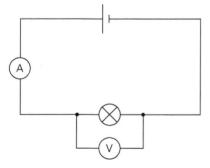

**Figure 2.6 Measuring the voltage across and the current through a lamp.**

## Resistance   REVISED

The amount of current flowing in a circuit depends on both the voltage applied and the resistance of the components. The resistance of a component can be worked out by using values from a voltmeter and ammeter.

$$resistance = \frac{voltage}{current}$$

$$R = \frac{V}{I}$$

> resistance, *R*, measured in ohms (Ω)
>
> voltage, *V*, measured in volts (V)
>
> current, *I*, measured in amperes or amps (A)

4.5 V is measured across the bulb in Figure 2.6. The ammeter reads 225 mA. What is the resistance?

Answer

$$R = \frac{V}{I}$$

$$R = \frac{4.5}{0.225}$$

$$R = 20\,\Omega$$

If two resistors are in series, the overall resistance is simply the total of the two values.

## Ohm's law

The resistance of any component in specific conditions can always be calculated with the measurements from a voltmeter and an ammeter. By increasing the voltage across a component, different readings of current can be recorded and a **current–voltage graph** (otherwise known as an *I–V* graph) can be drawn. The shape of the graph is different for different components.

If the graph is a straight line through the origin, the current is directly proportional to the voltage. This means the component has a fixed resistance in those conditions and is said to obey **Ohm's law**, or to be ohmic. The steeper the line, the lower the resistance.

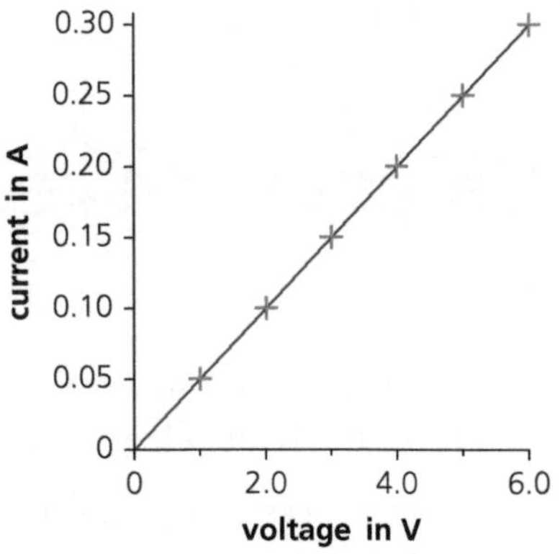

**Current–voltage graph:** A graph that is plotted with voltage along the horizontal or *x*-axis and current on the vertical or *y*-axis, and that shows the relationship between the two values. These graphs are sometimes called *I–V* graphs.

**Ohm's law:** This states that for some components, at constant temperature, the current through the component is proportional to the voltage across it. Such a component is said to be ohmic.

Ohm's law is often expressed as if it is being used to find the voltage:

$$V = I \times R$$

This arrangement is not particularly useful, but the equation can be rearranged depending on the data provided.

Using the graph in Figure 2.7, the resistor has a current of 0.2 A through it when the voltage across it is 4 V. What is the resistance?

Answer

$$R = \frac{V}{I}$$

$$R = \frac{4}{0.2}$$

$$R = 20\,\Omega$$

In most circuits we build in the classroom, currents will be small and measured in milliamps (mA); 1 A = 1000 mA.

Resistances may be large and may be measured in kilo-ohms (kΩ); 1 kΩ = 1000 Ω. Remember to convert to SI units before substituting into the equation.

**Figure 2.7** A current–voltage or *I–V* graph

It is easy to think that a steep line would mean a high resistance, but it is the opposite; a shallow line means a high voltage causes only a small current to flow.

voltage, *V*, measured in volts (V)

current, *I*, measured in amperes or amps (A)

resistance, *R*, measured in ohms (Ω)

Many components are not ohmic. Each different component has a different line on a current–voltage graph. A similar circuit can be used repeatedly, changing the component each time, to collect a set of data.

## The filament lamp

REVISED

At low voltages, when the temperature of the filament is low, so is the resistance. This means the $I$–$V$ line is steep, close to the origin and the current increases quickly. As the voltage increases the $I$–$V$ line becomes flatter and the current reaches a maximum. It is the increasing temperature that causes a higher resistance.

## The diode

REVISED

As its component symbol suggests, a diode only allows current to pass in one direction. This means that if the voltage is positive, the resistance is very low and the line is very steep. If the voltage is reversed, the resistance is so high that no current flows at all. A light-emitting diode (LED) behaves the same way, but also lights up to indicate when a current is flowing.

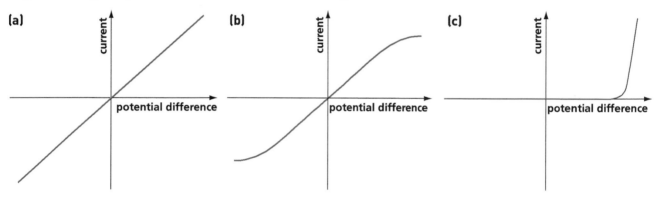

Figure 2.8 Current–voltage graphs show how components behave differently: (a) ohmic conductor, (b) filament lamp, (c) diode.

> **Exam tip**
>
> You should be able to recognise a component by the $I$–$V$ graph it produces and be able to explain how the shape relates to the resistance. It's worth learning the symbol and $I$–$V$ graph for each component as these often appear in exams.

## Changing resistance

REVISED

Some components vary in resistance depending on external conditions, such as temperature and light intensity. A **thermistor** behaves in the opposite way to most materials as it becomes a better conductor (has a lower resistance) when it is warm. A **light-dependent resistor** (LDR) is a poor conductor in the dark, but more current flows as more light shines on it – resistance decreases as light intensity increases.

> **Thermistor:** Component with a high resistance in cool or cold conditions and a low resistance when it is warmer. Different resistors have different ranges but the pattern (and shape) of the graph for each one is the same.
>
> **Light-dependent resistor (LDR):** Component with a high resistance in dark conditions and a low resistance when it is bright. Like thermistors, they are often used as sensors.

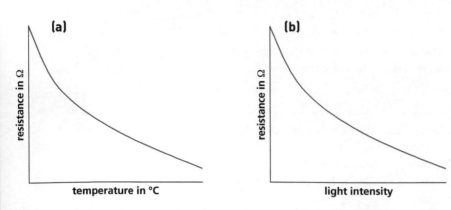

Figure 2.9 (a) Thermistors and (b) light-dependent resistors have changing resistance depending on external conditions.

## Now test yourself

15 Describe the difference in how an ammeter and a voltmeter are connected.
16 Give the symbol, unit and abbreviation for resistance.
17 A voltage of 12V is across a component, causing a current of 0.4A through it. What is the resistance?
18 A filament lamp lights up when current flows.
   (a) Draw the circuit symbol.
   (b) Sketch the shape of the $I$–$V$ graph (and label the axes) for this component.
19 What is the overall resistance of the circuit in Figure 2.10?

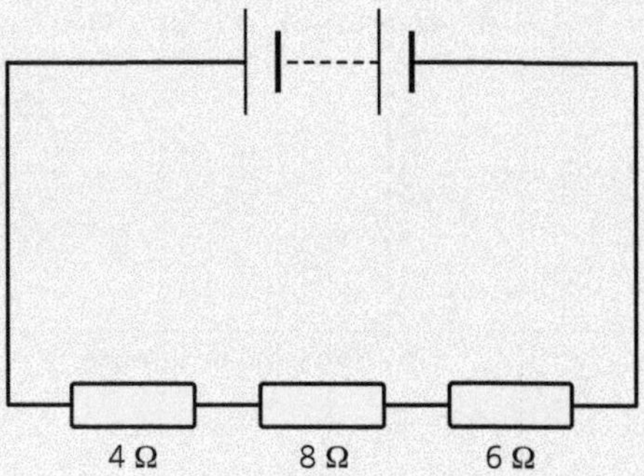

Answers on page 133

Figure 2.10 **Three resistors in series**

$4\,\Omega$    $8\,\Omega$    $6\,\Omega$

# Current, charge and voltage

## Current and charge

Current measures how much charge in **coulombs** passes a point in the circuit each second. The moving charges are electrons that travel along the wires from the negative terminal to the positive terminal. These electrons actually move quite slowly, but they all move at the same time so as soon as one electron leaves the cell another arrives.

To find out how much charge has moved past a point in the circuit we can use the following equation:

charge flow = current × time

$$Q = I \times t$$

> **Coulomb (C):** The SI unit of charge. Many electrons are needed to make up one coulomb.

charge, $Q$, measured in coulombs (C)

current, $I$, measured in amperes or amps (A)

time, $t$, measured in seconds (s)

An ammeter in a circuit gives a reading of 300 mA. How much charge moves through the ammeter in 5 minutes?

Answer

$Q = I \times t$

$Q = 0.3 \times 300$

$Q = 90\,C$

## Defining the volt

REVISED

Voltage measures how much energy is transferred by charges moving through a component in the circuit.

$$\text{voltage} = \frac{\text{energy}}{\text{charge}}$$

A reading of one volt means that one joule is transferred by each coulomb of charge. The most useful form of this equation is used to work out the energy transferred:

energy transferred = charge × voltage

$$E = Q \times V$$

In most situations the charge transferred ($Q$) won't be given. This means the first step is often to find the charge using current flowing and time.

energy, $E$, measured in joules (J)

charge, $Q$, measured in coulombs (C)

voltage, $V$, measured in volts (V)

A voltmeter connected across a resistor gives a reading of 0.8 V. A current of 0.04 A flows for 10 minutes. How much energy is transferred?

Answer

$Q = I \times t$

$Q = 0.04 \times 600$

$Q = 24\,C$

This means a charge of 24 coulombs is transferred in this time period. We can use this data to work out the energy as follows:

$E = Q \times V$

$E = 24 \times 0.8$

$E = 19.2\,J$

## Now test yourself

20 (a) What is the unit of charge? Include the abbreviation.
   (b) Define a volt in terms of energy and charge.
21 In a wire, what are the moving particles that transfer charge? Are they positively or negatively charged?
22 (a) A current of 43 mA flows for 2 minutes. How much charge was moved in this time?
   (b) If the voltage which caused this current was 2.4 V, how much energy was transferred?
23 What current will be measured if 36 coulombs pass through an ammeter in one minute?

Answers on page 133

# Current and voltage rules

## Current paths

If there is more than one loop in a circuit they are described as parallel to each other. The current before the **junction** is conserved, so that it is equal to the total of the currents flowing in each loop.

Current is a measure of how many charges move past a point each second. There are the same number of moving charges before a junction (when they are all together) and after (when they are on two or more loops); charges can't appear or disappear. This means the total of moving charges before and after the junction must be the same.

> **Junctions:** Places in a circuit where the current is divided between two loops, or where two loops join up again. A series circuit is one with no junctions and a single loop.

> **Exam tip**
> A voltmeter is always connected in parallel with a component.

### Example

What is the current at $I_2$ in Figure 2.11?

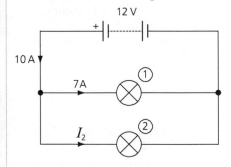

Figure 2.11 **Two lamps connected in parallel.**

Answer

The current before the junction is 10 A. This must be the same as the total through the two lamps.

$I_2 = 10 - 7$

$I_2 = 3\,A$

> **Short circuit:** When a connection is made but the current does not flow through higher resistance components. The current is often higher when this happens and can cause unexpected heating.

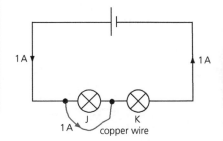

**Figure 2.12 When the short-circuit wire is connected, the current through lamp J decreases and the overall current, including through lamp K, increases.**

If a very low resistance path is connected as a parallel loop around other components, the current through the high resistance path may be almost zero. This is called a **short circuit** and can happen when wires bend or break inside a device.

## Adding the voltage

If components are connected to the power supply in parallel, they will have the same voltage across each of them. In Figure 2.11, both lamps have a voltage of 12 V across them.

If components are in series, the voltage across each of them will instead add up to the total supply voltage.

**Example**

What is $V_2$ in Figure 2.13?

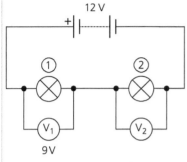

Figure 2.13 **Two lamps connected in series.**

Answer

The voltages across the two lamps must add up to the supply voltage of 12 V.

$$V_2 = 12 - 9$$

$$V_2 = 3\,\text{V}$$

**Exam tip**

In the lab, the numbers read from meters may not exactly match to the last decimal place. This is because of errors in the meters and the tiny voltage across the wires in the circuit. However, for any exam questions these minor variations in the numbers are removed.

## Uses of circuits

Parallel circuits are used when:
- all components in the circuit (for example, car lamps) need the same voltage to work properly
- components need to be controlled independently (for example domestic lighting)
- measuring the voltage with a voltmeter.

Series circuits are used when:
- one component needs to be controlled or protected by another, for example a switch, fuse or resistor
- each component needs a smaller voltage than can be easily supplied, for example fairy lights
- measuring the current with an ammeter.

## Cells and batteries

A battery is a term used for two or more cells in series. Each cell contains chemical substances that react together to cause a voltage between the terminals. All cells and batteries produce direct current.

If the cells are all connected the same way around (positive terminal to negative terminal) the voltages will add up. If they are reversed, the voltage may be reduced or at zero. For example, a 9V (square) battery contains six cells, each of which provides 1.5V.

## Now test yourself

24 (a) Are the lights in a house connected in series or parallel?
   (b) Why?
25 What is the overall current flowing in each circuit:
   (a) two components with 3A through each in series
   (b) three components in parallel with 2A through each
   (c) two components in parallel, one with 2A through it and the other with 1.5A?
26 Calculate the missing voltage value in each case:

(a)

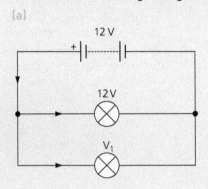

(b)

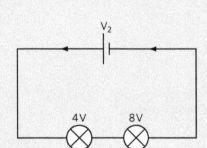

Figure 2.14

Answers on page 133

# Circuit calculations

A systematic approach to circuits is the only way to identify and calculate missing values. Consider which rules about current and voltage can be applied first, add in more numbers, and then think about what you can do next.

## Resistance, voltage and current calculations

REVISED

If any two values are identified, the third can be calculated. The three arrangements of the equation are:

$$R = \frac{V}{I} \qquad I = \frac{V}{R} \qquad V = I \times R$$

voltage, $V$, measured in volts (V)

current, $I$, measured in amperes or amps (A)

resistance, $R$, measured in ohms ($\Omega$)

### Typical mistake

Under pressure it is easy to substitute the first values you see into the equation. Check that you are using the values for the correct part of the circuit; you may wish to write on the circuit diagram to help.

## Power calculations

REVISED

If the voltage across a component and the current through it are known, the power transferred is given by:

$$P = I \times V$$

As before, check that the values used apply to the component.

## Now test yourself

27

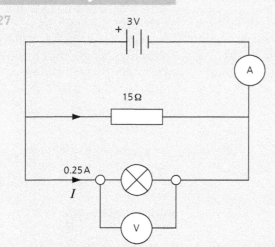

Figure 2.15 **Missing values in a circuit can be worked out.**

[a] If the cells are identical, what is the voltage of each?
[b] What is the voltage across the 15 Ω resistor?
[c] What is the current through the 15 Ω resistor?
[d] What is the resistance of the lamp?
[e] What is the reading on the ammeter?

Answers on page 133

Answers on page 133

> **Revision activity**
>
> All circuits provided in questions should follow these rules. Go through a worksheet, this guide or your textbook and each time check the rules for current, voltage and resistance. Where you start will be different each time but the aim is to get in the habit of checking if these calculations are possible: label each branch with the voltage. Add the current reading that would be shown by an ammeter. Where possible, work out the resistance.

# Electrostatics

## Electrical charge

Atoms are made of three particles: **protons**, **neutrons** and **electrons**. Protons and neutrons are found in the **nucleus** of the atom which has a positive charge as a result (protons have a positive chare, neutrons have no charge). The negatively charged electrons orbit the nucleus. The structure of the atom is covered in Chapter 7.

If an object gains more electrons than it had to begin with, it will become negatively charged. If it loses electrons, it will be positively charged. **Friction** between two insulating materials can lead both to become charged as one transfers electrons to the other.

Objects with the same sort of charge repel each other and objects with opposite charge attract each other. This is an example of a non-contact force (see page 5 for a reminder). A charged object can often pick up small items such as dust or small pieces of paper.

If a material is made up of many parts and each part is given the same charge, the parts will move as far from each other as possible, such as charged droplets of liquid that spread out from each other and someone's hairs separating out when they touch a Van de Graaff generator.

> **Protons**: Positively charged particles in the nucleus of an atom
>
> **Neutrons**: Uncharged particles in the nucleus of an atom
>
> **Electrons**: Negatively charged particles which orbit the atom and can be transferred by friction
>
> **Nucleus**: The centre of an atom, containing almost all the mass
>
> **Friction**: A force which resists the movement of two surfaces in contact

## Conductors

In metal solids and other electrical conductors, many of the electrons are free to move even though the atoms are fixed in place (see page 106 for more information). The negatively charged electrons are attracted to the positive terminal of a cell and are repelled by the negative terminal, so there is a continuous flow which is called current.

Some materials, such as plastics, hold an electrostatic charge well because the electrons are not free to move. This is why they are electrical **insulators**.

In some materials, the moving charges are **positive** or **negative ions** rather than electrons. These ions are atoms which have gained or lost electrons, often during chemical reactions. For example, ionic solutions conduct electricity.

> **Insulators**: Materials in which there are no charged particles like electrons that are free to move. They have a very high resistance. Commonly used insulators include rubber and plastic.
>
> **Positive ions**: Atoms that have lost electrons
>
> **Negative ions**: Atoms that have gained electrons

## Required practical

### Investigate how insulating materials can be charged by friction

#### Method

1 A gold-leaf electroscope was charged using a high-voltage supply. A note was made of the charge (positive or negative) of the electroscope.
2 A plastic rod was selected as an insulator and rubbed with a duster.
3 This insulator was held near the electroscope.
4 The gold leaf moved.
5 The results were recorded.
6 This experiment was repeated with different insulators and dusters.

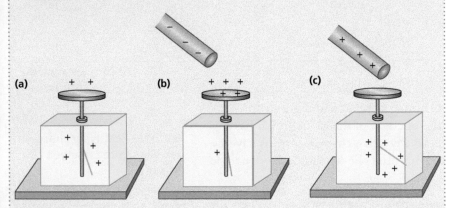

**Figure 2.16 Gold-leaf electroscope**

#### Analysis

The electroscope had an initial positive charge. When some insulators were held near the electroscope, the gold leaf fell, which showed that these insulators were negatively charged. When other insulators were held nearby, the leaf rose up; this showed that these insulators were positively charged.

> **Typical mistake**
>
> It is easy to get confused and think that losing something makes a negative ion. In fact, a negative ion gains negative electrons.

> ## Now test yourself
>
> 28 (a) Electrons are transferred to an insulator by friction. What sign charge will the insulator have?
>    (b) Two charged objects repel each other. What can you say about the sign of their charges?
> 29 (a) Copper atoms in a wire are fixed in place. Explain how charge moves when current flows.
>    (b) The atoms in the plastic around a cable have electrons. Why is the plastic an insulator?
>
> Answers on page 133
>
> TESTED

# Electrostatics at work

## Spark hazards

REVISED

Friction between insulators transfers electrons and builds charge up. If the difference in charge is high enough, a brief electrical current will flow which serves to balance the charges again. In air, this can be seen as a spark. If there are flammable materials present, such as petrol fumes when refilling vehicles, the spark may cause a fire.

To avoid this problem, cables are used that ground or 'earth' potentially charged vehicles when friction is likely because of fuelling. That means that any current will flow through the wire rather than the air and there is no spark, preventing a fire.

## Photocopying and printing

A photocopier works in several stages, as shown in Figure 2.17.

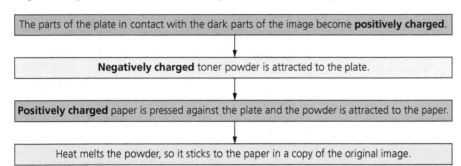

**Figure 2.17 How a photocopier works**

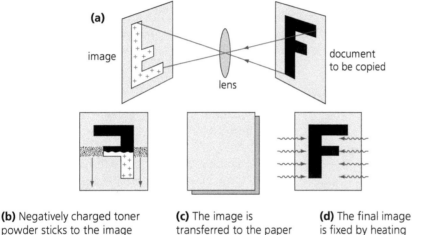

**(b)** Negatively charged toner powder sticks to the image

**(c)** The image is transferred to the paper

**(d)** The final image is fixed by heating

**Figure 2.18 Photocopying**

> **Typical mistake**
>
> You can lose easy marks if your written answers are incomplete. Make sure you link facts about opposite charges to attraction, and similar charges to repulsion.

> **Revision activity**
>
> These examples can be explained with a diagram and a few key words. For each one draw a clear diagram and add important words as labels. Look back at these to check that you can use the key terms to build an explanation. The aim is to return to the diagrams after a week, and this time recall the prompt words as well as the diagram.

A laser printer uses a process like that shown in Figure 2.17, which is why the paper is warm after printing. An inkjet printer also uses electrostatic charge, but in a different way. The ink droplets are charged, and then directed to the right place on the paper by a charged plate in the printer. This uses repulsion to change where the droplet goes.

## Now test yourself

TESTED

30 A refuelling vehicle is used to pump airplane fuel, which is an insulator, into a plane. A copper conductor is used during the refuelling process. Explain why the copper conductor is needed.
31 In a device called an electrostatic precipitator, soot particles pass near wires with a large negative charge.
   (a) What charge will the soot particles have now?
   (b) The soot must be collected on large plates around the wires. How should the plates be charged and why?

Answers on page 133

## Summary

- Voltage, $V$, measured in volts (V) causes charges to move.
- In a complete circuit these moving charges make a current, $I$, measured in amperes or amps (A). Small currents may be measured in milliamps (mA) with 1000 mA = 1 A.
- Energy is transferred when current flows. High current can cause heating and a high voltage can cause a spark. Both high currents and voltages can be dangerous.
- In the UK, mains voltage is 230 V. This is an alternating current (a.c.) which changes direction many times each second at a frequency of 50 hertz or 50 Hz.
- Cells and batteries provide a steady voltage which causes a direct current (d.c.) in one direction.
- Fuses, circuit breakers, earth wires and insulation are all safety devices that are used to reduce electrical hazards.
- Resistance, $R$, measures how hard it is to make a current flow in a material. It is measured in ohms ($\Omega$). Often kilo-ohms (k$\Omega$) are used, with 1000 $\Omega$ = 1 k$\Omega$.
- Power transferred by electricity, $P$, measured in watts (W) is calculated by:

$$P = I \times V$$

- To find the current flowing, if the power and voltage are known, this formula can be rearranged to:

$$I = \frac{P}{V}$$

- The cost of electricity will depend on how much energy, $E$, measured in joules (J) has been transferred. This is calculated by:

$$E = P \times t$$

- Circuit diagrams are drawn with straight lines and circuit symbols.
- In a series circuit all the components are in a single loop. Current is the same everywhere in the circuit and the voltages across each component add up to the overall voltage supplied by the source.
- If a circuit's voltage is increased or the resistance reduced, the current will be increased.
- To find the resistance of a component, the current through it is measured with an ammeter. The voltage across it is measured with a voltmeter.

$$R = \frac{V}{I}$$

- Ohm's law states that for some components, at constant temperature, the current through the component is proportional to the voltage across it. Such a component is said to be ohmic and a graph of current against voltage will be a straight line. Non-ohmic components such as filament lamps and diodes have their own characteristic $I$–$V$ graphs.
- A thermistor has high resistance when cold, which decreases with rising temperature; this is the opposite to most materials.
- A light dependent resistor (LDR) has high resistance in the dark and low resistance in bright light.
- Charge, $Q$, is measured in coulombs (C). In wires the moving charges are electrons. The amount of charge passing a point in the circuit each second is the current, $I$. So:

$$Q = I \times t$$

- The amount of energy, $E$, transferred by each coulomb of charge, $Q$, depends on the voltage, $V$:

$$E = Q \times V$$

- In a parallel circuit there will be more than one loop. Voltage across parallel loops will be the same and the total of the current in loops after a junction will be the same as the overall current before it.
- Friction can transfer electrons between touching surfaces. Objects that gain electrons have a negative charge. Objects which lose electrons have a positive charge.
- Objects with the opposite charge attract each other. Objects with the same charge repel each other.
- Objects that are highly charged can be dangerous as they can cause a brief current to flow. If this causes a spark then a fire can start.

# Exam practice

1 In most countries, electricity is supplied as an alternating current with a fixed voltage.
   (a) (i)  What is the voltage supplied by UK mains?
             A   12V        B   50V          C   120V              D   230V                          [1]
         (ii) State what current flows in a device which supplies 2.2kW using the UK mains.
              Give the unit in your answer.                                                          [4]
   (b) Identify a safety feature in a plug which breaks the circuit if the current is too high.      [1]
   (c) State which safety feature in a plug prevents a person who touches a metal case from
       receiving a shock.                                                                            [1]

2

A student builds a circuit and measures the current and voltage.

   (a) Assuming the cells are identical in the series circuit in the figure on the left, calculate the voltage supplied by each one.                                                                         [1]
   (b) The ammeter reads 30mA. Calculate the total resistance in the circuit.                        [3]
   (c) The voltmeter reads 2V. Calculate the voltage across the other lamp.                          [2]
   (d) The ammeter is now placed between the two lamps. State what the current reading will be.      [1]

3 A student is building a simple circuit that contains a diode.
   (a) Sketch the *I–V* graph for a diode.                                                           [3]
   (b) A student needs to collect data for a component to check it really is a diode. Describe the
       method needed to collect the data to create a graph so it can be compared with a textbook.    [3]
   (c) Explain, with reference to your sketch, why the circuit symbol for a diode includes an arrow. [2]

4 Electrical power is measured in watts while energy is measured in joules.
   (a) Calculate how much energy is transferred in 5 minutes by a laptop charger with a power
       rating of 6W.                                                                                 [3]
   (b) (i)  The current flowing in the laptop charger is 500mA. Calculate this in amperes.           [1]
         (ii) Calculate the voltage supplied to the device.                                          [3]

5 The circuit below is placed in a dark room. The current flowing is 0.06A.

   (a) (i)  Identify component X.                                                                    [1]
         (ii) Explain what will happen to the reading on the ammeter
              if the light in the room
              is turned on.                                                                          [1]
   (b) The voltage supplied to the component is 3V. Calculate the
       resistance in the dark.                                                                       [3]
   (c) Give an example of how this component might be used in
       everyday life.                                                                                [1]

6 A student builds a simple circuit with a lamp and a cell.
   (a) If a parallel loop is added with a resistor, explain what will happen to the brightness of the lamp. [1]
   (b) A current of 0.3A flows in that loop for 15 minutes. Calculate how much charge moves
       through the resistor.                                                                         [3]
   (c) Calculate how much energy is transferred by the resistor if the voltage across it is 2.4V.    [3]

7 A gold-leaf electroscope is charged positively by connecting to the positive terminal of a
  high-voltage supply.
   (a) Identify whether the electroscope has gained or lost electrons.                               [1]
   (b) A negatively charged rod is held near the cap. Explain why the gold leaf falls.               [2]
   (c) Explain what will happen to the gold leaf when a positively charged rod is held near the cap. [1]

## Answers and quick quizzes online

ONLINE

# 3 Waves

## Introducing waves

### Waves in solids and liquids

Many effects in physics can be described as **waves**. Ripples in water are examples of waves, as they transfer energy and information. The water does not move along with the ripple, although there may be a temporary change in vertical position. A floating object will not move along, just up and down. The exception to this is when an object, like a surfboard, is tilted to slide down from the peak to the trough as the wave moves.

> **Wave**: A disturbance that can transfer energy and information but not matter.

### Transverse and longitudinal waves

In a **transverse wave**, the material or medium is displaced (moved) and the wave travels at right angles (perpendicular) to the displacement. Energy is transferred with the wave motion.

> **Transverse wave**: This type of wave causes the material or medium to temporarily move at right angles to the direction of energy transfer. Light and ripples on water are good examples of this kind of wave.

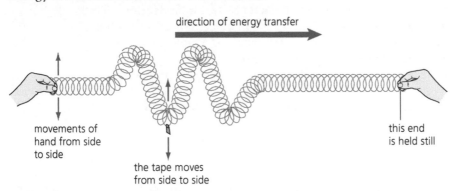

direction of energy transfer

movements of hand from side to side

the tape moves from side to side

this end is held still

**Figure 3.1 A transverse wave transfers energy at right angles to the displacement.**

A slinky spring on a flat surface can be used to demonstrate a transverse wave, if one end is moved from side to side. None of the material is moved permanently, but energy is transferred from one end to the other.

A slinky spring can also be used to demonstrate a **longitudinal wave**. Instead of a side-to-side movement, the end is pushed and pulled, towards the opposite end of the spring. Energy is also transferred with this wave motion.

> **Longitudinal wave**: This type of wave causes the material or medium to move temporarily in the same direction as the energy transfer. Sound is a good example of this kind of wave.

> **Exam tip**
>
> Avoid using a 'slinky' as an example of a transverse or longitudinal wave, because these springs can be examples of both types of waves. Instead use water waves, which are always transverse, and sound, which is always longitudinal.

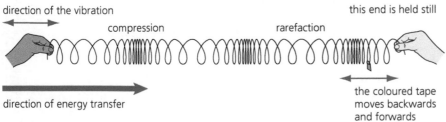

direction of the vibration

compression    rarefaction    this end is held still

direction of energy transfer

the coloured tape moves backwards and forwards

**Figure 3.2 A longitudinal wave transfers energy in the same direction as the displacement.**

# Describing waves

Both kinds of wave involve a temporary movement or change in the medium, sometimes called a displacement, and a permanent transfer of energy. The maximum displacement is the **amplitude**, $A$, and this depends on the amount of energy being transferred. For waves like sound and light, the amplitude is related to the loudness and brightness, respectively.

> **Amplitude, $A$**: The maximum displacement of a material. For a ripple on water, or a transverse wave on a slinky, it is the perpendicular height of the wave pulse, measured in metres (m), and is found by measuring how far the medium is displaced by the wave.

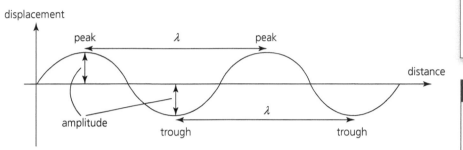

**Figure 3.3 Water ripples as seen from the side**

> **Typical mistake**
>
> When asked to find the amplitude, make sure you measure the height from the zero or equilibrium position. Students often measure from peak to trough, which is twice as far.

Each repeated motion causes one wave pulse; the distance between two pulses is called the **wavelength**. When part of a wave is seen or photographed, this can be measured. The easiest way is to measure between two peaks or two troughs. For a longitudinal wave, the distance is measured between two areas of compression or two areas of rarefaction.

> **Wavelength, $\lambda$**: The distance between two repeated patterns, measured in metres (m)

The amount of time taken for one complete wave pulse to pass a point is called the time period, $T$, of that wave. The number of complete pulses that pass any point in one second is called the frequency. These two quantities are related; as the frequency increases, the time period decreases. This is easy to show with a slinky, or ripples on water: if the source of the wave moves more often, the time for each wave pulse is less.

$$\text{frequency} = \frac{1}{\text{time period}}$$

$$f = \frac{1}{T}$$

You can change the subject of this equation if you need to work out the time period.

> frequency, $f$, measured in hertz (Hz)
>
> time period, $T$, measured in seconds (s)

> **Example**
>
> A motor causes a string to vibrate at 20 Hz. What is the time period of the wave?
>
> **Answer**
>
> $$T = \frac{1}{f}$$
>
> $$T = \frac{1}{20}$$
>
> $$T = 0.05\,\text{s}$$

## Wave speed and the wave equation

The **wave speed**, $v$, can be worked out like any speed: divide the distance travelled by the time taken (see page 1 for more examples). But this value connects other important wave quantities as well, giving us the wave equation.

wave speed = frequency × wavelength

$$v = f \times \lambda$$

> **Wave speed, $v$:** The speed at which the wave moves through the medium. It is measured in metres per second (m/s). For electromagnetic waves in a vacuum the value is always the same, $3 \times 10^8$ m/s, and is referred to as $c$.

### Example

Students measure water ripples on a pond caused by a model boat. Four ripples arrive each second and the distance between each one is 0.02 m. Calculate the wave speed for these ripples.

Answer

$$v = f \times \lambda$$

$$v = 4 \times 0.02$$

$$v = 0.08\,\text{m/s}$$

> wave speed, $v$, measured in metres per second (m/s)
>
> frequency, $f$, measured in hertz (Hz)
>
> wavelength, $\lambda$, measured in metres (m)

### Now test yourself

1 (a) What is the unit of frequency?
  (b) Calculate the frequency of a wave that has a time period of 0.02 seconds.
2 Is sound an example of a transverse or longitudinal wave?
3 In air, a sound wave has a wave speed of 330 m/s. What is the frequency if the wavelength is measured as 0.2 m?

Answers on pages 133–4

> **Typical mistake**
>
> It is often easier to measure wavelengths in centimetres, but this must be converted to metres before calculating wave speed. Pause to check the final answer makes sense to avoid losing marks.

> **Wavefront:** A line where all the water affected by a wave rises or falls together. A straight beam vibrating in water of uniform depth causes a straight wavefront.

# Ripple tanks

Ripples on water are good examples of transverse waves. The water rises and falls as the wave transfers energy along the surface of the water. To make them easier to see, a ripple tank is set up so that the peaks and troughs cast shadows. When the vibrating source is a long beam that is dipped in and out of the water, each pulse forms a long line called a **wavefront**.

The lines of each wavefront are perpendicular to the movement of the wave. The gap between them is the wavelength. When the wave meets a barrier, or a boundary between materials, the angle at which the wave meets the barrier changes what happens next. A ripple tank can be used to investigate what happens for both barriers and boundaries.

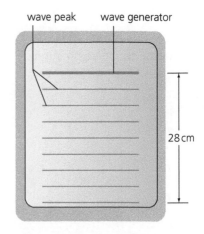

**Figure 3.4 The wave generator causes a series of ripples, each with a visible peak.**

# Reflection

If a rigid barrier is placed in the path of a wavefront, **reflection** happens. To make it easier to see the difference in waves before and after reflection, the barrier in a wave tank should be placed at an angle.

> **Reflection**: A change in the direction of a wave because of a barrier. The angle of incidence (before the barrier) always equals the angle of reflection (after the barrier).

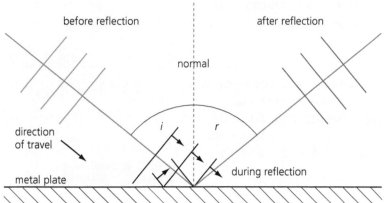

**Figure 3.5 The angle of incidence *i* is equal to the angle of reflection *r*, measured from the normal.**

The **angle of incidence** and the **angle of reflection** are always measured from the **normal**, an imaginary line which is perpendicular to the surface. The angle of incidence, $i$, is the angle between the arriving wave and the normal. The angle of reflection, $r$, is the angle between the reflected wave (moving away from the surface) and the normal. These angles are equal to each other, which can be expressed as:

$i = r$

> **Angle of incidence, $i$**: The angle between the incident ray and the normal. It is usually measured with a protractor.
>
> **Angle of reflection, $r$**: The angle between the reflected ray and the normal
>
> **Normal**: A line placed at right angles to a surface or boundary

> **Typical mistake**
>
> When describing a method to investigate the direction of travel, make sure you specify that angles are measured from the normal. Students lose marks if they state or imply that the angles are measured between the surface and the wave.

Like all waves, light and sound can be reflected too (and a reflected sound is called an echo).

# Refraction

If the wave meets a boundary between different materials instead of a barrier, there may be **refraction** as well as reflection. This is where the direction of the wave motion changes as it crosses the boundary. This is seen in a ripple tank when the depth changes.

> **Refraction**: A change in the direction of a wave when it crosses a boundary between materials. The amount of refraction depends on the difference in properties of the two materials and the angle of incidence.

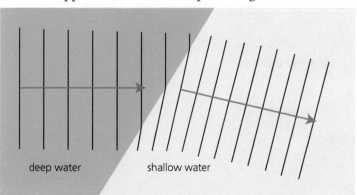

**Figure 3.6 The wave changes direction at the boundary between deep and shallow water.**

Refraction happens because the speed of the wave is different before and after the boundary. This can be seen by comparing the wavelength of ripples in deep and shallow water where the frequency does not change. This change in speed is easiest to see if the incident wave meets the boundary along the normal, as there is no refraction to confuse the observations. All waves are refracted when they meet a boundary at an angle, although the effects may sometimes be small.

## Doppler effect

If the source of a wave and the observer are stationary in comparison to each other, the detected frequency and wavelength are the same as at the source. If they are moving in comparison to each other, there will be a difference between measurements taken at the source compared to measurements taken at the observer. This is called the Doppler effect:

● If the source is getting closer, the observed frequency will be higher and the wavelength will be lower.
● If the source is moving away, the observed frequency will be lower and the wavelength will be higher.

The faster the source is moving, the bigger the change in frequency and wavelength. This effect will be familiar to most people when thinking of sound waves. For example, an approaching car will seem to have a higher pitch or frequency as it approaches, and it will seem to have a lower pitch once it is moving away. You can see page 57 for more information on the characteristics of sound waves. The Doppler effect is observed with all types of waves, although it is not always easy to measure.

> **Revision activity**
>
> Write the name of each variable used in this topic on an index card. On the back, write the symbol, unit and definition. Choose one to review and test yourself on each day. When you can recall each one separately, start testing yourself on three randomly chosen cards.

## Now test yourself

4 Explain why the normal line for a surface is important when investigating waves.
5 What is the wavelength of the ripples in Figure 3.4?
6 As a motorbike approaches an observer standing on a pavement, and then passes, the engine noise seems to get louder then quieter again. What other change will the observer notice?
7

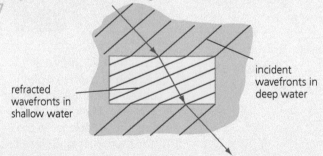

refracted wavefronts in shallow water

incident wavefronts in deep water

Figure 3.7 **Ripples changing direction as they pass from deep to shallow water.**

(a) How has the wavelength changed as the ripples pass into the shallow water?
(b) What name is used for the change of direction?

Answers on page 133

# Electromagnetic waves

## The electromagnetic spectrum

REVISED

Although they seem very different, radio waves and visible light are both **electromagnetic waves**. They behave differently because they have different wavelengths and frequencies, but they make up a complete and continuous spectrum. All electromagnetic waves travel at the same speed in a vacuum. This is $3 \times 10^8$ m/s and is often called the 'speed of light'.

> **Electromagnetic waves**: Transverse waves which can travel through a vacuum as well as matter. Energy is transferred by changing electric and magnetic fields.

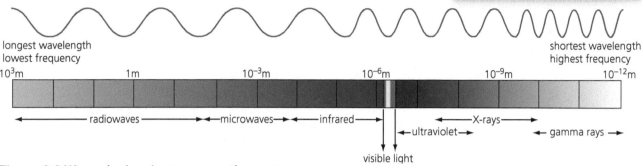

Figure 3.8 **Waves in the electromagnetic spectrum**

## Electric and magnetic fields

REVISED

Instead of the temporary displacement of matter, electromagnetic waves transfer energy because of changes in electric and magnetic fields. These changes are at right angles to the direction in which the wave is travelling.

## Wave speed

REVISED

Electromagnetic waves all travel at $3 \times 10^8$ m/s in free space (in a vacuum). Through other materials, including air, the speed is lower than this. The wave equation applies to electromagnetic waves, but because the speed is always the same it is often given as $c$ rather than $v$.

> **Exam tip**
>
> All waves, including electromagnetic ones like light, obey the same wave equation. A different symbol is often used for the speed of electromagnetic waves to remind us that the speed in a vacuum is the same for all of them. In a vacuum $c$ is a constant, not a variable.

> **Example**
>
> An electromagnetic wave has a wavelength of 2 cm. What is the frequency?
>
> Answer
>
> $c = f \times \lambda$
>
> $f = \dfrac{c}{\lambda}$
>
> $f = \dfrac{3 \times 10^8}{0.02}$
>
> $f = 1.5 \times 10^{10}$ Hz

> **Exam tip**
>
> In exam questions, the frequency of an electromagnetic wave might be expressed in kilohertz (kHz), megahertz (MHz) or gigahertz (GHz).
>
> - 1 kHz = 1000 Hz
> - 1 MHz = 1000 kHz = 1 000 000 Hz = $1 \times 10^6$ Hz
> - 1 GHz = 1000 MHz or 1 000 000 000 Hz = $1 \times 10^9$ Hz

# Bands of the electromagnetic spectrum

Although the basic characteristics of electromagnetic waves are the same, the details vary as the frequency and wavelengths change. The electromagnetic spectrum is continuous, but it is usually divided into seven convenient bands, depending on how the characteristics vary from a scientific point of view. In reality, the difference between a low-wavelength radio wave and a high-wavelength microwave is actually very small.

## Radio waves

The longest wavelength and lowest frequency electromagnetic waves are called radio waves and these are used for broadcasting and communications. Those with a lower wavelength have a shorter range. When absorbed, for example by an aerial, they produce an alternating current with the same frequency as the wave.

## Microwaves

With a wavelength measured in centimetres or millimetres, microwaves have several uses. As well as being used for shorter range communication transmissions, for example between mobile phone masts and satellites, they also cause heating when absorbed. This is how a microwave oven works, as the waves penetrate food to cook the centre of it, instead of just heating the surface. Metal shielding on microwave ovens prevents the potential internal heating of our own body tissues.

## Infrared waves

Any object that has a higher temperature than its environment, for example a room heater, emits infrared waves. These waves are sometimes described as thermal radiation and they can burn the skin. Humans can detect infrared waves as warmth, for example from the Sun, but the waves are not visible to our eyes. Infrared photography is used to measure temperature and these images are often given false colours to show which areas are hotter. Night-vision cameras can use this to distinguish warm objects, such as animals, from a cold background at night.

## Visible light waves

Often referred to simply as 'light', these are the waves which can be detected by the human eye. Different parts of the visible spectrum have different wavelengths, from $7 \times 10^{-7}$ m (which is seen as red light) to $4 \times 10^{-7}$ m (which is seen as violet light). As well as photography, visible light is used to send information along optical fibres.

## Ultraviolet waves

With wavelengths less than violet light and as low as $10^{-9}$ m (1 nm), ultraviolet waves are not visible to humans but can damage the eyes or skin. Hot objects above 4000 °C, for example the Sun, can emit ultraviolet waves. Damage to our surface cells (sunburn) is caused by the ultraviolet waves which are part of normal sunlight, even on a cold day. Prolonged exposure can cause skin cancer, which is why sunblock and protective clothing are recommended if people are spending long periods of time outside. If eyes are affected, there is also a risk of blindness.

Some materials absorb ultraviolet waves and emit visible light. This is called fluorescence, and is the basis for many lights. An electric current produces ultraviolet light, which is absorbed by a special coating that

> **Exam tip**
>
> Because of the scale being used, the wavelengths of electromagnetic waves between infrared and ultraviolet are often given in nanometres (nm). In these units the visible spectrum ranges from 700 nm for red light to 400 nm for violet light. When you see these units in exam questions, make sure you convert to standard units (metres).

emits visible light instead. Dyes and inks that behave in this way are sometimes used for security marking: the labels are invisible until an ultraviolet source is directed at them.

## X-rays

Electromagnetic waves with a wavelength of around $10^{-10}$ m pass through many soft materials, including human muscle and organs, but are stopped by hard materials such as bone or metal. These waves can cause mutations in living tissue, but are useful in medicine because they show damage to bones or teeth inside the body. They are also used in industrial situations to look inside objects that cannot easily be taken apart.

## Gamma rays

The shortest wavelength electromagnetic waves are emitted from the nuclei of atoms (see page 108 for more information). These waves are highly penetrating and can cause mutations in cells which may lead to cancer (even more so than X-rays and ultraviolet waves). Lead shielding is used to protect those who regularly work with or near X-ray and gamma ray sources. Gamma rays are used to sterilise food and medical equipment as they can kill micro-organisms.

> **Revision activity**
>
> This topic includes a lot of facts which need to be memorised, but this will be easier if you can link them to everyday uses or implications. Aim to find these links and include them in your notes. To improve recall, you could write seven sentences that make up a memorable story, one for each wave band, and include a fact for each one.

> **Revision activity**
>
> This topic requires you to be fluent with the maths needed to link frequency and wavelength for very large or small numbers; it is easy to mix up the number of zeros involved. Start by practising the uses of the equation in each possible rearrangement (with speed, wavelength or frequency as the subject). Part of doing the maths well involves using your calculator correctly, so you are confident with the use of standard form and know where brackets are needed.

## Now test yourself

TESTED

8 Which parts of the electromagnetic spectrum
   (a) are used for communications
   (b) cause mutations in living cells
   (c) are located between microwaves and ultraviolet waves?
9 An ultraviolet wave has a wavelength of 350 nm.
   (a) What is this in metres?
   (b) What is the frequency?
10 From your knowledge of the visible spectrum, which has a higher frequency: orange or blue light? You don't need to give the values.
11 Explain how a person might get sunburnt while skiing on a very cold day.

Answers on page 134

# Reflection

## Reflection of light rays

REVISED

The beams of light produced from a **ray box** show where the waves are travelling. Most ray boxes contain a bright filament lamp and so they can become very hot in use. The position of each beam is shown on a diagram as a straight line, called a ray, with an arrow to show the direction of travel. Ray diagrams are always drawn with a ruler.

> **Ray box:** A box containing a bright lamp, with a narrow slit so that the light from the lamp inside escapes as a narrow beam.

If the beam from a light box meets a smooth, shiny surface it will be reflected. Light is reflected from rough surfaces too, but because the surface angles are all different, there is no clear image.

When investigating the reflection of light, you need to know that the **incident ray** strikes the mirror at the angle of incidence, $i$, measured between the ray and the normal. This angle is always the same as the angle of reflection, $r$, measured between the **reflected ray** and the normal. This relationship can be expressed as: $i = r$.

> **Incident ray:** The ray which meets a surface or boundary
>
> **Reflected ray:** The ray that travels away from a shiny surface like a mirror

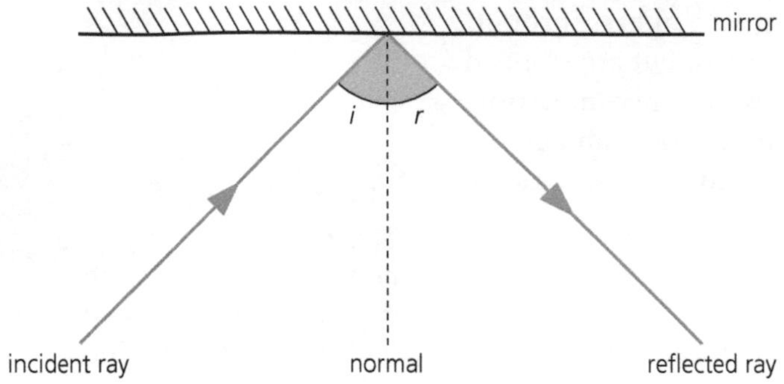

**Figure 3.9 Reflection of light by a mirror**

The incident ray, the reflected ray and the normal always lie in the same plane. This means they can all be plotted on the same flat surface.

## Now test yourself

TESTED

12 Draw a ray diagram for a beam with an angle of incidence of 30° which is reflected by a mirror.
13 What is the angle between the normal and the surface of a mirror called?
14 The angle between an incident ray and a mirror is 15°. What is the angle of reflection?

Answers on page 134

# Refraction

## Refraction of light rays

REVISED

When a wave crosses a boundary between two materials it will change speed. If it meets the boundary at an angle, this will cause the wave to change direction. This change in direction is called refraction. The **angle of refraction** where light crosses the boundary between materials depends on the properties of the materials and the angle of incidence.

> **Angle of refraction:** The angle between the refracted ray and the normal

### Required practical

### Investigate the refraction of light, using rectangular blocks, semi-circular blocks and triangular prisms

An experiment was conducted to measure the angle of refraction for light that crosses from air into a glass **prism** at different angles of incidence.

#### Method

1 The ray box was set up and adjusted to produce a narrow beam.
2 The block was positioned on the paper so the incident ray met the surface.
3 The incident ray and the position of the refracted ray leaving the far side of the block were both marked with a pencil and labelled.
4 The ray box was moved to change the angle of incidence. Six pairs of incident and refracted rays were marked and labelled. ⇨

> **Prism:** A shaped transparent object, usually glass or plastic, which is polished so light is refracted and reflected by the surfaces

**5** A protractor was used to measure the angle of incidence and angle of refraction for each pair of rays.

**6** Different students investigated the angle of refraction with other shapes including a triangular prism and a semi-circular glass block.

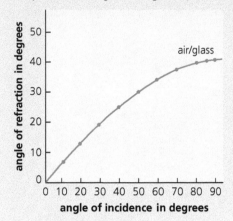

**Figure 3.10 Angle of refraction plotted against angle of incidence for a rectangular block.**

## Analysis

The graph shows that as the angle of incidence increases, the angle of refraction increases. The angle of refraction is always less than the angle of incidence when the ray enters a denser material like glass. The difference between the two angles increases as the angle of incidence approaches 90°.

The results show that light is refracted towards the normal when it goes into a glass block. It is refracted away from the normal when it leaves the glass. This is also true for the semi-circular blocks and triangular prisms.

# Refractive index

REVISED

As well as affecting the angle of incidence, the materials involved also make a difference to the angle of refraction. We call this the **refractive index**, $n$, of the material.

$$\text{refractive index} = \frac{\text{speed of light in air}}{\text{speed of light in the material}}$$

**Refractive index:** The ratio of the speed of light in air to the speed of light in a material. It has no units.

The speed of light is very high and cannot easily be measured. Because of this, it is much easier to find the refractive index by calculating the values of the angle of incidence and the angle of refraction of a beam entering the material.

The relationship is not just between the angles. Instead, a mathematical function is used that you will remember from work on triangles. The sine of the angle of incidence, $\sin i$, is divided by the sine of the angle of refraction, $\sin r$:

$$n = \frac{\sin i}{\sin r}$$

refractive index, $n$, a ratio of speeds with no unit

angle of incidence, $i$, measured in degrees (°)

angle of refraction, $r$, measured in degrees (°)

## Required practical

### Investigate the refractive index of glass, using a glass block

The procedure outlined in the previous Required practical can be used to investigate the refractive index of a glass block.

### Method

1 Columns were added to the results table for sin *i* and sin *r*.
2 A graph was plotted of sin *i* on the *y*-axis against sin *r* on the *x*-axis.
3 The gradient of the line of best fit was measured.

| Angle of incidence/° | Angle of refraction/° | sin *i* | sin *r* | sin *i*/sin *r* |
|---|---|---|---|---|
| 0 | 0 | 0 | 0 | |
| 23 | 15 | 0.39 | 0.26 | 1.50 |
| 34 | 22 | 0.56 | 0.37 | 1.51 |
| 48 | 30 | 0.74 | 0.50 | 1.48 |
| 59 | 35 | 0.86 | 0.57 | 1.51 |
| 80 | 41 | 0.98 | 0.66 | 1.48 |

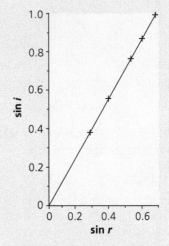

**Figure 3.11 The gradient of the graph is the refractive index of glass.**

### Analysis

The graph shows that the refractive index of the glass of the rectangular block is approximately 1.5. This can be checked by choosing any pair of values, as shown in the following example, and calculating *n*. The points on the graph are close to the line of best fit, showing that the errors in the results are small.

### Example

Using the third line from the table, an angle of incidence of 34° causes an angle of refraction of 22°.

#### Answer

We can use this information to calculate the refractive index of the glass:

$$n = \frac{\sin i}{\sin r}$$

$$n = \frac{\sin 34}{\sin 22}$$

$$n = 1.51$$

## Refraction by prisms

REVISED

Different shaped objects, called prisms, each have their own patterns for refraction. However, the rules at each surface are the same: light is refracted towards the normal when it enters the prism, and away from the normal when it leaves. The difference between the original and final direction of the light will depend on the shape of the prism and the angle of incidence.

## Now test yourself

15 For a glass block, will the angle of refraction be more or less than the angle of incidence?
16 For practicals that test light refraction, what safety precaution should be taken?
17 Light crosses from air into glass, which has a refractive index of 1.5, at 30°. What is the angle of refraction?
18 Diamond has a refractive index of 2.42. Calculate the speed of light in diamond.

Answers on page 134

# Total internal reflection

Each time light meets a glass boundary it may be reflected or refracted. We often pay attention to just one or the other, but both may happen at the same time. If a semi-circular glass block is set up so light travelling inside meets the surface at different angles, it can be seen that above a specific angle all of the light is reflected and none is refracted. This behaviour is called **total internal reflection**.

> **Total internal reflection:** When light, travelling inside a material, meets the surface but all the light is reflected inside, with none being refracted out of the material

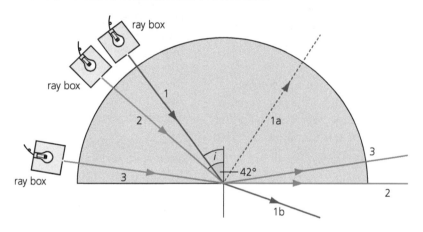

**Figure 3.12 If the incident angle is below 42° light is reflected (1a) and refracted (1b). At 42° the light is refracted to travel along the surface of the prism (2). Above 42° all the light is reflected (3).**

## Critical angle and refractive index

In Figure 3.12, the maximum angle of incidence before total internal reflection is 42°. This is a typical **critical angle** for glass. Different materials have different values. For example, in water the critical angle is around 49°.

The critical angle, $c$, for a material can be calculated if the refractive index is known. You do not need to know anything about the material on the other side of the boundary because there is no refraction. The formula to calculate this is:

> **Critical angle, $c$:** The critical angle for a material is the maximum angle of incidence for which there is any refraction. Above this value all light will be reflected within the material.

$$\sin c = \frac{1}{n}$$

critical angle, $c$, measured in degrees (°)

refractive index, $n$, a ratio with no units

Diamond has a refractive index of 2.42. What is the critical angle?

Answer

$$\sin c = \frac{1}{n}$$

$$\sin c = \frac{1}{2.42}$$

$$\sin c = 0.41$$

$$c = \sin^{-1}(0.41)$$

$$c = 24° \text{ (to nearest degree)}$$

# Total internal reflection in prisms and optical fibres

REVISED

If light meets an ordinary mirror at an angle, there will be several beams because of the reflection and refraction at the surface. These multiple beams can make the image seem blurred. For situations where clear images are important (for example in cameras, optical instruments and periscopes) two triangular prisms are used. The triangular prisms are arranged so that light crosses the boundaries between air and glass along the normal. This means there is no refraction, as all of the light is reflected instead.

Optical fibres are long and flexible strands of glass. The materials for the two layers in the fibres are chosen so that light travelling in the inner core is always totally reflected. Optical fibres have several important uses, including communications (where visible or infrared light carries information as coded pulses) and in **endoscopes** used by medical professionals.

**Endoscope**: Bundles of optical fibres that can be connected to a camera. One end is placed inside a patient's body or another object, receives light through some fibres, and an image can be seen through the other end.

## Now test yourself

TESTED

19 (a) Define the critical angle for a material.
   (b) What is the critical angle for glass?
20 The critical angle for water is 49°. What is its refractive index?
21 Diamond has a relatively low critical angle. What effect does this have on its appearance and why?

**Revision activity**

Create a concept map linking together key ideas, examples and applications of refraction or reflection. When you are finished, cover up a quarter of the page and practise recalling the hidden ideas. Each time you repeat this, cover up a different part of the map.

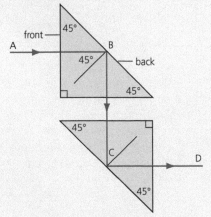

Figure 3.13 **Two prisms used to make a periscope.**

22 Explain why, in the diagram of a periscope in Figure 3.13, there is no refraction shown.

**Answers on page 134**

# Sound waves

Sound travels as longitudinal waves which can be reflected and refracted. Although we often think of sound as travelling through gases like air, these waves can also transfer energy through solids and liquids.

## The ear

REVISED

The ear is made up of several parts, many of which lie inside the body. For example, inside the body there is a thin surface called the ear drum, which moves when air next to it is compressed and decompressed by a sound wave. As well as the ear drum, tiny bones and fluid contained within shaped chambers are caused to vibrate, which eventually causes nerves to transmit an electrical signal to the brain. These signals are decoded by the brain to recognise sounds.

## Making and hearing sounds

REVISED

The source of a sound is always something which vibrates. This vibration causes a medium, such as air, to vibrate too. The vibrating movement back and forth creates a longitudinal wave made up of regions where the molecules are pushed closer together (**compressions**) and pulled further apart (**decompressions**). The distance between two compressions in a wave is the wavelength.

A wave may be detected by an instrument such as an ear drum or a microphone. Human ears can detect vibrations in the range from 20 to 20 000 Hz, although the upper limit is usually reduced by accumulated damage over our lifetimes. Waves with a frequency above 20 000 Hz are called ultrasound and they can be detected by some animals, but not humans.

> **Compressions**: Regions where the temporary displacement of a medium has pushed particles closer together
>
> **Decompressions**: Regions where the temporary displacement of a medium has moved molecules further apart. These are also sometimes called rarefactions.

> **Typical mistake**
>
> Do not use the term 'expansion' to describe regions where the wave has caused the particles to move further apart. Decompression or rarefaction makes it clear you understand the movement is a wave behaviour, rather than something caused by heating.

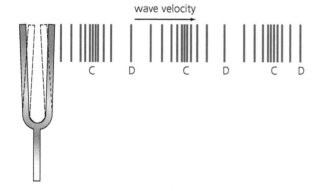

**Figure 3.14 The compressions (C) and decompressions (D) make up a sound wave.**

## Speed of sound

REVISED

When people talk about the 'speed of sound', they usually mean the speed of sound 'in air', but even this value varies. Sound travels quicker in liquids and much more quickly in solids. Finding the speed of sound, just as with any other speed, means measuring the distance travelled and the time taken (see page 1 for more information). If the distance is small, then the time will need to be measured very carefully.

$$\text{speed} = \frac{\text{distance travelled}}{\text{time taken}}$$

## Required practical

### Investigate the speed of sound in air

Two different methods are used to allow the calculated values to be compared.

### Method 1

1 A loudspeaker was connected to a signal generator which produced short bursts of sound.
2 Two microphones, separated by a distance, *d*, were placed near the loudspeaker.
3 Each microphone was connected to a dual beam oscilloscope so the two detected signals were displayed at the same time.
4 The scale on the oscilloscope screen was used to calculate the time taken, *t*, for a wave pulse to move between the two microphones.

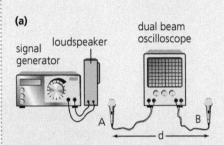

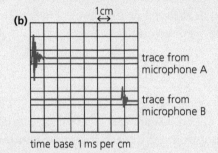

**Figure 3.15 The screen shows the same pulse detected by each microphone.**

For microphones 220 cm apart, the time taken was measured as 6.5 ms.

### Method 2

1 A student stood 40 m in front of a tall building and clapped their hands.
2 Each time they heard the echo, they clapped again.
3 A second student used a stopwatch to record the time for 10 echoes.
4 This was repeated three times so that a mean time for 10 echoes could be calculated.

It is important to remember that the distance travelled for each echo was 80 m, from the student to the wall and back again.

| Trial number | Time in seconds |
|---|---|
| 1 | 2.4 |
| 2 | 2.8 |
| 3 | 2.3 |
| Mean | 2.5 |

### Analysis

| Data from Method 1 | Data from Method 2 |
|---|---|
| $speed = \dfrac{d}{t}$ | $speed = \dfrac{d}{t}$ |
| $speed = \dfrac{2.2}{0.0065}$ | $speed = \dfrac{80}{0.25}$ |
| speed = 338 m/s | speed = 320 m/s |

### Conclusion

The values are similar, but not identical, and close to the accepted value for the speed of sound in air of 330 m/s.

## Now test yourself

TESTED

23 (a) Taking the speed of sound in air as 330 m/s, what is the wavelength for a wave with a frequency of 40 Hz?
   (b) How many times will the ear drum vibrate in a minute for this sound?
24 Describe how you would model a sound wave using a slinky.

Answers on page 134

**Typical mistake**

Students lose marks if they say a change in amplitude or frequency makes a detector such as the ear drum 'vibrate more'. Be specific: does it vibrate faster (frequency) or vibrate further back and forth (amplitude)?

# Loudness and pitch

The amplitude of a wave is the maximum displacement of the medium, and is related to the energy carried by the wave. A sound wave with higher amplitude of vibration is therefore louder, and temporarily moves the ear drum a larger distance. If the amplitude is large enough the ear drum may be damaged.

The frequency of a wave is the number of complete cycles or patterns each second, measured in hertz (Hz). A sound wave with a high frequency is squeaky, whereas a sound wave with a low frequency is called 'deep' or 'low'. The term **pitch** is used to describe how we experience this property of a sound. The vibration of the source of a sound, for example a guitar string, determines the frequency of the wave.

**Pitch**: The pitch of a sound is how we detect the frequency of the wave. High-pitched sounds, like a bat squeaking, are high-frequency waves. Low-pitched sounds, like an elephant's call, are low-frequency waves.

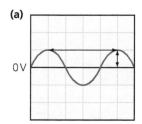 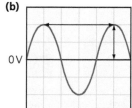 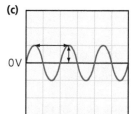 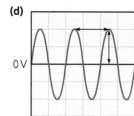

Figure 3.16 Graphs to show (a) low amplitude, low frequency, (b) high amplitude, low frequency, (c) low amplitude, high frequency, (d) high amplitude, high frequency.

## Using an oscilloscope to measure frequency

REVISED

Because it is hard to see a pattern in the compressions and decompressions of a sound wave, an **oscilloscope** is often used to help us. The vibrations of a sound wave cause the microphone to move backwards and forwards.

The amount of displacement is measured (see page 97 for more information) and plotted vertically on the oscilloscope trace against the time, just like you would on a graph. The controls on the oscilloscope set the scale of the graph being drawn.

## Required practical

# Investigate the frequency of a sound wave using an oscilloscope

An experiment was conducted to find the frequency of a sound wave. An oscilloscope was used to measure the time period of the sound and this was used to calculate the frequency.

### Method

1 The microphone was connected to the input of the oscilloscope and it was switched on.
2 The $y$-offset control was adjusted so that the visible line was across the middle of the oscilloscope screen.
3 With the source of the sound close to the microphone, the $y$-gain control was adjusted so that one complete waveform, peak to trough, was visible vertically on the screen.
4 The $x$-gain control was adjusted so that one complete waveform was visible horizontally on the screen. It was important to check that it was the simplest possible waveform that could be displayed.
5 Using the setting of the $x$-gain control and the length of the waveform on the oscilloscope screen, as shown below, the time period was measured as 8 ms.

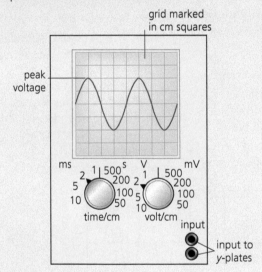

**Figure 3.17 Each horizontal division on the screen is equal to 2 ms.**

### Analysis

$$\text{frequency} = \frac{1}{\text{time period}}$$

$$\text{frequency} = \frac{1}{0.008}$$

$$\text{frequency} = 125 \, \text{Hz}$$

---

**Oscilloscope**: An instrument that measures changing voltage over time and displays this on the screen as a line. When the sensor connected to an oscilloscope is a microphone, the changing voltage represents the changing displacement of the medium caused by a sound wave.

**Exam tip**

It is important to remember that no matter how the information is displayed, a sound wave involves **longitudinal** displacement. An oscilloscope connected to a microphone makes a trace that might look like a transverse wave, but in fact it is a graph showing displacement by a longitudinal wave.

**Typical mistake**

For this experiment only one microphone is needed as the wave itself is being investigated. Only if two waves are being compared, or if looking at the time taken for the wave to move between two positions, will you need to use two microphones.

## Now test yourself

25  A student suggests that the sound of a jet engine has a high frequency because it is loud. What mistake have they made?

26  Which control on an oscilloscope will need to be adjusted if the waveform is too wide to fit on the screen?

27  A microphone is connected to an oscilloscope and used to display a sound wave from a musical instrument. How will the graph shown vary when
    (a)  the instrument is played more loudly
    (b)  a higher note is played?

28  A musical note at 600 Hz is played into a microphone. How wide will the waveform on the oscilloscope screen be if the scale is set to 5 ms per centimetre?

Answers on page 134

### Revision activity

How many of the variables from this topic can you recall without checking your notes at all? Include the mathematical relationships between them in your definitions. Try explaining the concepts with only words, or with only a diagram. Which is easier?

## Summary

- Waves transfer energy without permanent movement of a material.
- Longitudinal waves like sound have a temporary displacement in the same direction as the transfer of energy.
- Transverse waves like ripples on water have a temporary displacement at right angles to the transfer of energy.
- Amplitude, $A$, is the maximum displacement of a wave, related to the energy carried by it.
- Wavelength, $\lambda$, is the distance measured between two identical parts of a repeated waveform, measured in metres (m).
- Time period, $T$, is the time taken for one complete waveform to pass a given point, measured in seconds (s).
- Frequency, $f$, is the number of complete waveforms that pass a given point each second, measured in hertz (Hz).

$$\text{frequency} = \frac{1}{\text{time period}}$$

$$f = \frac{1}{T}$$

- wave speed = frequency × wavelength

$$v = f \times \lambda$$

- There is a change in the observed frequency of a wave when the source is moving relative to the observer. This is called the Doppler effect.
- All waves reflect and refract at a boundary or surface.
- All electromagnetic waves are transverse, travelling at the same high speed in a vacuum: $c = 3 \times 10^8$ m/s.

- The electromagnetic spectrum, in order of increasing frequency (and decreasing wavelength) is radio waves, microwaves, infrared (IR), visible light, ultraviolet (UV), X-rays, gamma rays. There are specific uses and dangers for each band.
- Light, the visible part of the electromagnetic spectrum, can be reflected and refracted. When reflected, the angle of incidence, $i$, equals the angle of reflection, $r$. Angles are always measured from the normal, a line perpendicular to the point on the surface where the light arrives or leaves the material.
- Refraction can be observed when light appears to bend at a boundary between materials. The refractive index, $n$, can be calculated using:

$$n = \frac{\sin i}{\sin r}$$

- All materials have a critical angle, $c$, above which all incident beams are reflected rather than refracted. This is called total internal reflection and is important for the use of optical fibres.

$$\sin c = \frac{1}{n}$$

- Sound is a longitudinal wave which can be reflected and refracted. The audible frequency for humans is between 20 Hz and 20 000 Hz. Above this range it is called ultrasound.
- The amplitude of a sound wave is related to the loudness. Loud sounds are caused by larger vibrations.
- The frequency of a sound wave is related to the pitch. High-pitched sounds are caused by vibrations at a higher frequency.

# Exam practice

1 A sound wave is displayed on an oscilloscope screen as shown in the figure below.

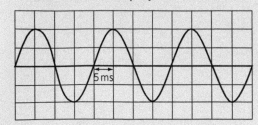

(a) Which device is connected to the oscilloscope to detect the sound wave? [1]

| A | ammeter | C | microphone |
|---|---------|---|------------|
| B | voltmeter | D | loudspeaker |

(b) Using the scale given, determine the time period of the wave. [1]

(c) Calculate the frequency of the detected sound. Give the unit. [4]

2 A student is investigating reflection of light using a mirror.

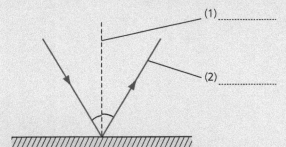

(a) Identify the labels which should be added to features 1 and 2. [2]

(b) For one reading the angle of incidence is 30°. Predict the angle of reflection. [1]

(c) The student states that light is a transverse wave. Explain what this means. [2]

3 The visible part of the electromagnetic spectrum covers the range from red to violet light.

(a) Identify which colour of light has the highest frequency. [1]

(b) Red light has a wavelength of 700 nm.

   (i) State the value of this wavelength in metres. [1]

   (ii) Using $c = 3 \times 10^8$ m/s, calculate the frequency in hertz. [3]

4 A campaigner suggests that mobile phone signals cause brain cancer.

(a) (i) Which two parts of the electromagnetic spectrum are used for mobile phone signals? [2]

| A | visible light | C | microwaves |
|---|---------------|---|------------|
| B | infrared | D | radio waves |

   (ii) X-rays are known to cause ionisation in cells. State another part of the electromagnetic spectrum which can also cause ionisation. [1]

(b) (i) Describe how the properties of X-rays are useful in medical investigations. [2]

   (ii) Explain how medical staff protect themselves from X-rays. [1]

(c) Explain why it is unlikely that the campaigner is correct. [2]

5 Light enters the cornea at an angle. The cornea is part of the eye.

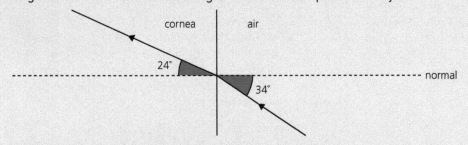

(a) Identify the angle of incidence shown in the diagram. [1]

(b) (i) Calculate the refractive index for the material of the cornea. [3]

   (ii) Explain why refractive index has no unit. [1]

(c) A synthetic material is produced to repair a damaged cornea. During a test researchers measure the critical angle as 44°.

   (i) Define the critical angle. [2]

   (ii) Use the measured value to find the refractive index of the synthetic material. [3]

   (iii) Will this material be useful as a replacement for a damaged cornea? [1]

## Answers and quick quizzes online

ONLINE

# 4 Energy resources and energy transfer

## Energy

### Energy stores and systems

When a **system** changes, the energy associated with different stores also changes. The changes in a system are measurable (for example the height a ball is lifted from the floor).

> **System**: An object or a group of objects that interact with each other

It is important to understand which changes in the system relate to which energy stores:

- Gravitational: When an object is lifted or dropped, the energy in the gravitational store changes. This is sometimes called the gravitational potential energy (see page 68).
- Kinetic: A moving object has energy in a kinetic store. This is sometimes called the kinetic energy (see page 68).
- Thermal: Any object with a temperature above absolute zero has energy in a thermal store (see page 82).
- Chemical: The energy in the chemical store changes during chemical reactions, including those that occur in living things. This is often called chemical potential energy (see page 70).
- Magnetic: There is a change in the magnetic store when the distance between two magnetic poles increases or decreases (see page 88).
- Electrostatic: There is a change in the electrostatic store when the distance between two electric charges increases or decreases (see page 37).
- Elastic: Materials that are temporarily stretched or compressed have increased the amount of energy in an elastic store (see page 10).
- Nuclear: When a nucleus changes there will be an increase or decrease in the nuclear store.

Physicists use equations to describe the energy changes, although not all of the stores have equations that you need to know at this level.

When a system changes, energy is transferred between stores. For example, if objects are thrown up and then fall down, the kinetic and gravitational stores are affected. Another example is during endothermic and exothermic reactions when there are transfers between chemical and thermal stores. When any system changes at least one store will be decreased and at least one will be increased. Sometimes more than two stores are affected.

The principle of conservation of energy states that the amount of energy always remains the same. It can be transferred between stores but never destroyed or created.

Energy, $E$, is measured in joules (J). One joule is a comparatively small amount of energy in most situations, so it is often helpful to use kilojoules (kJ) and megajoules (MJ).

$$1\,kJ = 1000\,J = 1 \times 10^3\,J$$

$$1\,MJ = 1000\,kJ = 1000000\,J = 1 \times 10^6\,J$$

> **Typical mistake**
>
> Students can lose marks when they say that energy has been 'lost'. Usually what they mean is that it is no longer easy to measure, or that it is no longer in a useful store.

Many processes can transfer energy between stores:
- waves, like sound and light
- heating by particles
- electrical current.

These are all processes that transfer energy, not stores. A chemical reaction is another kind of process which transfers energy, as is a mechanical transfer that involves forces.

## Useful and wasted energy transfers

Some processes are termed 'useful' because they transfer energy to a store that we plan to use. Often, processes that occur in a system transfer energy to stores that we can't use. This is sometimes described as 'wasted' energy. In most situations where the energy is transferred in a way which is not useful, it ends up transferred to the thermal store of the surroundings. This process is called **dissipation**.

> **Dissipation**: The transfer of energy to the surroundings, usually to a thermal store, so it is no longer useful

## Efficiency

Some machines or devices waste more energy than others. The energy that is transferred in useful ways (or to useful stores), when considered as a proportion of the energy supplied, is called the **efficiency**. Efficiency is worked out as a fraction but is often expressed as a decimal or a percentage.

$$\text{efficiency} = \frac{\text{useful energy output}}{\text{total energy input}} \times 100\%$$

> **Efficiency**: The proportion of supplied energy that is transferred to a useful store

> **Exam tip**
>
> Efficiency can be expressed as a fraction, decimal or percentage – in each case it means the same thing. Remember that it can never be more than 1 (or 100%).

> **Example**
>
> A winch is supplied with 12 kJ. It transfers 10.5 kJ to the gravitational store, which is useful. What is the efficiency?
>
> Answer
>
> $$\text{efficiency} = \frac{\text{useful energy output}}{\text{total energy input}} \times 100\%$$
>
> $$\text{efficiency} = \frac{10.5}{12} \times 100\%$$
>
> $$\text{efficiency} = 87.5\%$$

One way to represent the efficiency of a process is to draw a **Sankey diagram** with arrows for each transfer of energy. The width of the arrow is proportional to the amount of energy transferred in each way, or to each different store.

> **Sankey diagram**: A diagram that shows how much energy is transferred to different stores or through different processes

> **Revision activity**
>
> Consider a different real-life process every evening and start by identifying one store that has lost energy and at least one which has gained energy. Once you are confident about the variables involved with the start and end points, you will be ready to choose the right equations in questions that require a mathematical approach.

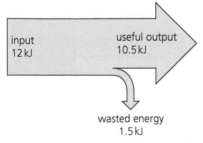

**Figure 4.1 A Sankey diagram for the winch given in the example. The width of each arrow represents the amount of energy.**

## Now test yourself

1 For each system, state the store that energy is being transferred to:
   (a) a box lifted from the floor to a table
   (b) water in a kettle being heated by the element
   (c) the string of a bow pulled back by an archer
   (d) a ball bearing rolling down a ramp and speeding up.
2 Give the following values in joules:
   (a) 28 kJ
   (b) 4.3 MJ
   (c) 0.91 kJ
   (d) 76.5 MJ.
3 In one minute, a light bulb receives 240 J and transfers all this energy to the thermal store of the room. 60 J is transferred as visible light, while the remainder is transferred by different kinds of heating. What is the efficiency?

Answers on page 134

# Conduction and convection

Energy is normally transferred from hotter objects to cooler ones, but how this transfer happens depends on the materials involved and the surroundings. Heat transfer between particles in a material is called conduction.

## Insulators and conductors

Materials that allow heat to be transferred through them quickly, from particle to particle, are called conductors. Metals are very good conductors, because they have free electrons that can move throughout the material. Most building materials, such as brick and concrete, are also reasonable thermal conductors (although they are not conductors of electricity).

If the transfer of heat is slow, the material is an insulator. Air, plastics and wood are described as insulators. Mammals and birds also have layers of insulating tissue that reduce the thermal transfer of energy to the surroundings. A vacuum – a region with no particles at all, like space – prevents thermal transfer by conduction and is used in engineering solutions to reduce heat transfer.

> **Exam tip**
>
> If an exam question involves both heat and electricity, be specific about the kind of conductor or insulator you are describing. For example, brick is a fairly good thermal conductor but a bad electrical conductor.

## Required practical

### Investigate thermal energy transfer by conduction

An experiment was conducted to compare the thermal conduction properties of four metals.

#### Method

1 Four metal samples were placed so they could be heated equally by a Bunsen burner flame.
2 A droplet of water was placed in the hollow at the end of each sample.
3 The Bunsen burner was lit and a stopwatch used to measure the time taken for the water to boil on each metal.

#### Analysis

Copper was the best conductor, with the water boiling after only 10 seconds. Aluminium was only slightly worse at 16 seconds. The water on brass took 36 seconds, and iron was the worst conductor tested with a time of 54 seconds.

# Convection currents

When materials are heated they usually expand. If a fluid (liquid or gas) is heated, the expanded region becomes less dense and will rise in comparison to the unheated regions. This process is called convection and it transfers energy because the particles in the material move.

Whenever air is heated it rises, while cooler unheated air falls. This pattern of moving air is called a convection current. This type of current can happen within a small space such as a room, as well as in the large space of the Earth's atmosphere, where it accounts for the direction of wind from the sea in the day, and towards the sea at night. This is because of the different heating rates of the land and sea.

If the air being heated is trapped, for example in layers of clothing, the thermal transfer of energy is reduced. So although air is an insulator in terms of conduction, the air must be prevented from moving if convection is to be minimised. Convection needs particles that are free to move, so it cannot happen in solids or in a vacuum.

### Typical mistake

The everyday description of convection is that 'heat rises'. A much better description is that 'heated substances rise' as it shows that you understand heated particles are moving.

## Required practical

### Investigate thermal energy transfer by convection

An experiment was carried out to demonstrate the convection currents in water.

### Method

1 A large flask was filled with water and placed on a tripod with gauze. This was left to settle for several minutes.
2 A crystal of potassium permanganate (VII) was dropped into the centre of the flask.
3 The centre of the flask was heated slowly and the path of the dissolved potassium permanganate (VII) was observed.

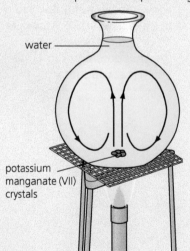

water

potassium manganate (VII) crystals

**Figure 4.2 The path of the dissolved potassium permanganate (VII) was observed.**

### Analysis

The observed paths of potassium permanganate (VII) colour showed that the heated water rose, spread out along the top surface, then fell as the water cooled. The water rose when it was warmer and so became less dense than the water around it, and fell when it was cooler.

### Exam tip

Remember that the convection currents are **caused** by the heating, but are made **visible** because of the potassium permanganate (VII). These currents are there even when they cannot be seen because it is the water particles that are moving.

## Unwanted energy transfer

REVISED

Materials that trap air in bubbles or pockets are good at reducing both conduction and convection. These materials are used to avoid large energy transfers where they are not wanted, for example in homes and other buildings:

- Chimneys are often placed within buildings so that the warm air heats the walls of upstairs rooms before entering the atmosphere.
- External walls often consist of two layers of brick with foam sealed between. Although brick is a relatively good thermal conductor, the energy transfer is reduced by the thickness of material and the trapped air. The area of the walls, as well as the temperature difference between inside and outside, also affects the rate of heat transfer but these factors are harder to change.
- Materials like carpets and loft insulation, which trap air to reduce both conduction and convection, are effective ways to reduce energy transfer to the outside of a building. Thicker materials trap more air, so are better insulators.
- Double glazing also traps air, or sometimes other gases like argon, in between two layers of glass. The insulating material here is the gas rather than the glass, which is a relatively good thermal conductor.

> **Exam tip**
>
> In most cases, the air that is trapped within a material is the effective insulator. Remember that because it is trapped, both convection and conduction are reduced.

### Now test yourself

TESTED

4 On a hot day at the beach the air above the land is heated more quickly than the air above the water. Sketch the movement of the air that results.
5 Why will there be a faster transfer of energy to the outside of a house on a winter's day compared to the energy transfer in summer?
6 Which properties make a metal like copper such a good thermal conductor?

Answers on page 135

# Radiation

As well as being transferred through contact between particles (conduction) and movement of particles (convection), energy can also be transferred without particles. This type of energy transfer is called thermal radiation, which is another way to describe infrared waves (see page 48 for more information).

## Absorbers and emitters

REVISED

Bright and shiny materials, which have polished surfaces and pale colours, tend to reflect thermal radiation instead of absorbing it. These types of materials are also bad emitters, so when they are hot they transfer energy away relatively slowly.

Materials with dark, matt surfaces are good absorbers and emitters of thermal radiation. They do not reflect much of the thermal radiation that falls on their surfaces.

> **Typical mistake**
>
> It's easy to mix up conduction and radiation when shiny metal materials are involved. Metal materials are good conductors, but poor absorbers and emitters of radiation because they are shiny. If it is important that heat is transferred by both conduction and radiation, the metal is usually painted a dark colour.

> **Exam tip**
>
> If in doubt about what types of surfaces reflect or absorb radiation, think about colours of clothing that you wear outside on warm days. Dark colours warm up more quickly, as they absorb thermal radiation from the Sun at an increased speed. White clothes, even though the same amount of radiation falls on them, absorb less and so do not warm up as quickly.

The higher the temperature difference between an object and its environment, the more thermal radiation is emitted and the faster energy is transferred to the thermal store of the surroundings. There is also a change in the wavelength of the emitted radiation (see page 47 for more details).

## Required practical

### Investigate thermal energy transfer by radiation

An experiment was conducted to compare how quickly thermal radiation was absorbed by different surfaces.

### Method

1 Two sheets of aluminium, one left shiny and the other painted matt black, were prepared by sticking a marble on each surface with wax, which was allowed to set.
2 The two sheets were placed at an equal distance from a heater.
3 The heater was switched on and the time taken for each marble to fall off was recorded.

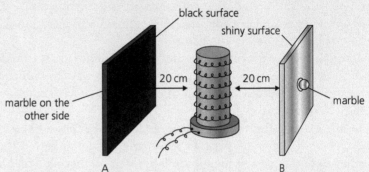

**Figure 4.3 The time taken for each marble to fall off was recorded.**

### Results

The marble on the sheet with the painted side facing the heater dropped off first, after 34 seconds. The marble on the sheet with the shiny side facing the heater dropped off after 53 seconds.

The results show that dark matt surfaces absorb thermal radiation faster than shiny surfaces.

## Now test yourself

TESTED

7 In a practical to compare the absorption of thermal radiation by different surfaces, as well as the two surfaces used in the practical described earlier on this page, a third aluminium sheet painted light grey is used. Predict how long it will take the marble to fall off. Give reasons for your answer.
8 Which part of the electromagnetic spectrum is described as 'thermal radiation'?
9 A student lays two plastic sheets on snow that has settled on a bright but cold day. One sheet is white, the other black. Predict what will happen over the next few hours. Give reasons for your answer.
10 Why are the cooling tubes on the back of a refrigerator painted black?

Answers on page 135

### Revision activity

Anything designed to keep things warm or cool will rely on the same basic principles. Pick any two examples and compare them; you should be able to explain similarities and differences. For example, both fleece jackets and loft insulation trap air which reduces energy losses by conduction and convection.

# What is work?

## Work done

REVISED

Physicists say that **work** is done whenever a force is applied to move an object over a distance. The amount of work done depends on both the force needed and the distance moved.

Work is measured in joules (J). This is the same as the unit used for energy, because when work is done energy is always transferred. Work is not a store, but rather a way that energy is transferred between stores.

**Work**: The energy transferred when a force is applied to move an object over a distance

The equation to calculate work done is:

work done = force × distance

$$W = F \times d$$

When one newton is applied to move an object by one metre, this is the equivalent of one joule of work being done.

| work done, $W$, measured in joules (J) |
| --- |
| force, $F$, measured in newtons (N) |
| distance, $d$, measured in metres (m) |

### Example

A student drags a box along a table with a newtonmeter. The reading given shows the force applied is 15 N, and the box is moved 4 m. What is the work done?

Answer

$W = F \times d$

$W = 15 \times 4$

$W = 60\,J$

When work is done, energy is transferred between stores. In the previous Example, energy has been transferred from the chemical store associated with the student's muscles to the thermal store of the table and box, with intermediate steps involving the kinetic stores, because they are warmed by the friction caused by movement.

Work is only done when a force is applied to cause movement over a distance. Energy can be transferred without movement, for example when holding an object so it does not fall down; in this example work is not done and the energy is transferred in other ways.

## Work and energy transfers

REVISED

Energy is transferred between stores as a result of physical processes. These processes often involve forces, fields and waves. It can be helpful to describe how these processes happen. Energy can be transferred by:
- a force moving something through a distance (mechanical working)
- a current flowing in a material or circuit (electrical working)
- conduction and convection (heating by particles)
- different kinds of waves or radiation, including both light and sound.

Sometimes, although we know that energy is transferred, it may not be obvious which of these transfer types is happening. Chemical reactions, including those in biological processes like photosynthesis, transfer energy through the movement of electrons, but it can be hard to recognise this as a form of electrical working. It is important to remember that these are useful descriptions rather than definitions.

If the starting point of an energy transfer is a chemical store, this usually means a fuel (such as petrol or digested food) is used up. Many of the processes described above eventually lead to heating of the environment, and both light and sound cause a temperature rise even though it is much too small to measure in a school science lab.

**Now test yourself**

TESTED

11 Rearrange the formula for work done so that the following values can be worked out:
   (a) distance moved
   (b) force applied.
12 If 80 J is transferred from the chemical store of a battery when a model car moves 4 m, what force is being applied?
13 If 240 J is transferred when an object that has a weight of 32 N is lifted, what is the increase in height?
14 A battery is connected to a filament lamp. Describe the processes transferring energy:
   (a) from the battery to the filament
   (b) from the filament to the room.

Answers on page 135

# Calculating energy

Energy stores can be described mathematically, but you don't need to know the maths for all of them at this level. Apart from gravitational potential energy (GPE) and kinetic energy (KE), the only other formula for energy you will need is the one for specific heat capacity, which is explained on page 82.

## Gravitational potential energy

REVISED

When an object is lifted up, the work done is calculated using the weight of the object and the increase in height. The weight of the object depends on both the mass, $m$, and the strength of gravity, $g$, which on Earth is 10 N/kg (see page 8 for more information). The work done to lift an object up is transferred to the gravitational store and is sometimes called the gravitational potential energy (GPE).

> mass, $m$, measured in kilograms (kg)
>
> gravitational field strength, $g$, measured in newtons per kilogram (N/kg). On Earth, $g = 10$ N/kg.
>
> height, $h$, measured in metres (m)

gravitational potential energy = mass × gravitational field strength × height

$$GPE = m \times g \times h$$

**Example**

A mountaineer, who has a mass of 60 kg, climbs 20 m up a cliff. What is the increase in their GPE?

Answer

$GPE = m \times g \times h$

$GPE = 60 \times 10 \times 20$

$GPE = 12\,000$ J or 12 kJ

## Kinetic energy

REVISED

A moving object is said to have energy in a kinetic store. This is sometimes described as having KE. The amount of energy depends on how fast the object is travelling and how much mass it has, but the relationship is not straightforward.

> mass, $m$, measured in kilograms (kg)
>
> velocity, $v$, measured in metres per second (m/s)

kinetic energy $= \frac{1}{2} \times$ mass $\times$ speed$^2$

$$KE = \frac{1}{2} \times m \times v^2$$

**Example**

A skateboarder, who has a total mass of 45 kg, is moving at 3 m/s. What is their kinetic energy?

Answer

$$KE = \frac{1}{2} \times m \times v^2$$

$$KE = \frac{1}{2} \times 45 \times 3^2$$

$$KE = \frac{1}{2} \times 45 \times 9$$

$$KE = 202.5\,J$$

Because the velocity is squared in this formula, it only takes a small increase in velocity to make a big difference to the energy in the kinetic store. For example, doubling the velocity increases the energy by a factor of four.

## Using equations to calculate energy changes

REVISED

The principle of conservation of energy states that energy cannot be created or destroyed, only transferred from one store to another. In other words, the total energy before a process must be the same as the total energy after. This means that, ignoring any energy transferred to thermal stores, the energy decrease in a gravitational store must be the same as the energy increase in the kinetic store (or the other way around).

$$m \times g \times h = \frac{1}{2} \times m \times v^2$$

In other cases, a change in the kinetic store can be worked out by using the formula for work done.

$$F \times d = \frac{1}{2} \times m \times v^2$$

If you are asked to calculate a value using either expression, start by identifying all the values given and rearrange the formula, depending on what you are being asked to work out.

## Power

REVISED

The rate at which energy is transferred is called the power (see page 25 for more information on power calculations in electrical circuits). If the same amount of work is done in less time, the power is greater.

$$power = \frac{work\ done}{time\ taken}$$

$$P = \frac{W}{t}$$

power, $P$, measured in watts (W)

work done, $W$, measured in joules (J)

time, $t$, measured in seconds (s)

If work is done very quickly, the power may be measured in kilowatts (kW) or megawatts (MW).

$$1\,kW = 1000\,W = 1 \times 10^3\,W$$

$$1\,MW = 1000\,kW = 1000\,000\,W = 1 \times 10^6\,W$$

### Example

An industrial crane lifts a mass of 500 kg by 24 m. This takes 30 seconds. What is the power of the crane during this action?

Answer

$$P = \frac{W}{t}$$

$$P = \frac{(500 \times 10 \times 24)}{30}$$

$$P = \frac{120\,000}{30}$$

$$P = 4000\,W \text{ or } 4\,kW$$

## Now test yourself

15  A tennis ball, with a mass of 60 g, is moving at 20 m/s. What is its kinetic energy?
16  How much work is done when a 75 kg student walks upstairs to the top floor in school, where there is an increase in height of 12 m?
17  A winch can supply power at 40 W. How long will it take to lift a mass of 10 kg by 8 m?
18  After slipping, a rock climber, who has a mass of 85 kg, falls 12 m before their safety line stops them. Ignoring air resistance, how fast would they be travelling at that point?
19  How much energy is transferred in 5 minutes if the power of a device is 2.7 kW?

Answers on page 135

# Energy resources and electricity generation

## Electricity generation and energy transfers

Mains electricity can be generated in several different ways. Most of these methods of generation rely on the movement of a magnet relative to a coil of wire. This movement causes or induces a current in the wire, as described further on page 98. Anything which can generate electricity is called an energy resource, which can be **renewable** or **non-renewable**.

Describing the energy transfers for each energy resource can help us to understand how they work. Most transfers involve a kinetic store because a generator has a moving magnet in a coil of wire.

**Renewable energy:** An energy resource that will not run out for the foreseeable future, for example, solar power

**Non-renewable energy:** An energy resource that has a limited supply (even if it is expected to last many years), such as fossil fuels and uranium

## Fossil fuels

Coal, oil and gas are described as fossil fuels because they are formed over millions of years by the decay of living things in anaerobic conditions. Different situations create a range of fossil fuels and so they may be solids, liquids or gases. These fuels (chemical stores) are burnt to heat water to steam (thermal store). The steam created causes turbines to spin around (a kinetic store), which does mechanical work on the generator to produce an electric current.

| Advantages | Disadvantages |
|---|---|
| Fossil fuels have been readily available for many years and the power stations that use them are reliable. Gas-fired power stations can be turned on and off quickly and can provide a lot of power. | All fossil fuels are non-renewable and as the fuels become scarce their costs increase. Burning fossil fuels also produces carbon dioxide, a greenhouse gas which contributes to climate change. Coal produces other gases which can further damage our health and the environment. |

## Nuclear power

REVISED

In a nuclear power station, the fuel used is uranium or plutonium. These elements are not burnt, but instead transfer energy from the nuclear store during nuclear fission (for more information see page 116). As in a fossil-fuel power station, this turns water to steam (thermal store) which turns a turbine (kinetic store) and then a generator. Instead of a chemical store the starting point is the nuclear store of the atom, which transfers energy to the kinetic store of fast-moving particles, which then causes the heating.

| Advantages | Disadvantages |
|---|---|
| Although non-renewable, the available uranium will last for much longer than remaining fossil fuels. Far more energy is produced from a kilogram of uranium than any type of chemical fuel. No pollutant gases are produced in normal operation. | The small amount of radioactive waste produced in this method is dangerous for hundreds of years, and must be stored safely. It also takes a long time to start and stop the power station running. Accidents are rare but can be serious, causing contamination of the local and wider area. |

## Wind power

REVISED

Air that is moving naturally in the atmosphere can be used to turn a turbine. Heating by the Sun transfers energy to the kinetic store of the air, and some of this energy is then transferred to the turbine blade and generator. Wind turbines can be placed on land or off-shore.

| Advantages | Disadvantages |
|---|---|
| No pollutant gases are produced and the running cost is low because no fuel is needed. It is a renewable resource. | Wind speed is highly variable and so the amount of electricity generated is unpredictable. This means that wind turbines must be used either to charge up a battery or in combination with a more consistent supply. The turbines can also be noisy (sound pollution) and some people complain that they are ugly (visual pollution). |

## Water power

REVISED

Moving water has been used by humans to do work for centuries. If the flowing water in a river or the sea is used to turn a turbine, electricity can be generated.

A dam can be built to trap rain water high up on a river's course in an artificial lake called a reservoir. By allowing this water to flow through a turbine, energy is transferred from the gravitational store of the water to the kinetic store of the turbine. This is called hydroelectric power and works best in countries with lots of rain and high hills.

Tidal power relies on the rise and fall of the sea water in river estuaries. As the water flows in and out of the estuary at each low and high tide it passes through turbines.

| Advantages | Disadvantages |
|---|---|
| Hydroelectric and tidal power work without producing any pollutant gases. Tidal power works because of tides, which are both reliable and predictable. Hydroelectric power is reliable as long as there is plenty of rain. Once built, running costs are low as no fuel is needed. | Building hydroelectric and tidal power stations can be very expensive initially and disruptive to the environment in the long term. Their creation may involve flooding areas where people and animals live. Neither method works consistently through the whole day.<br><br>Potentially, wave power using waves moving on the surface of the sea could be used to drive turbines. Unfortunately, these waves are caused by the wind and so far it has proved much more effective to use wind turbines instead. |

## Geothermal

REVISED

In volcanic areas there may be hot rocks close to the Earth's surface. When water is pumped down to these rocks steam is produced, which can turn turbines to generate electricity. The rocks are hot because of nuclear processes that occur naturally deep in the Earth's crust, so the starting point is the nuclear store of the materials below the surface.

| Advantages | Disadvantages |
|---|---|
| There are no waste products and geothermal power is a renewable resource. In suitable areas, the running costs are low and power generation is constant. | Only a few countries, such as Iceland and New Zealand, have areas that are suitable for geothermal power on a large scale. Setting up a power station is challenging, expensive and can be dangerous because of the volcanic gases that are released. |

## Solar cells

REVISED

A solar cell generates electricity when special materials absorb light. No turbines or other moving parts are involved. Solar cells on a building may reduce the need for electricity from the National Grid. If sunshine is reliable throughout the year, solar 'farms' with many linked solar cells can be effective.

| Advantages | Disadvantages |
|---|---|
| Running costs are very low and there are no waste products. Solar power is renewable and can be very useful in remote places, especially in locations with long days and where the weather is predictable. | The installation costs are high and the systems must include batteries to store the energy generated for night-time use. If the weather is variable, the output is unreliable so may need to be combined with more consistent resources. |

## Solar heating systems

REVISED

One way to reduce electricity consumption is to use the sunlight as a method of heating. Normally this means using sunlight to heat water in pipes, which reduces the amount of heating by electricity or gas.

| Advantages | Disadvantages |
|---|---|
| A solar heating system is much cheaper to build than a solar cell system and it reduces both electricity/fuel use and heating costs. There are no waste products and it relies on a renewable resource. | In many regions the amount of sunlight is unreliable and the system cannot be used at night. It can only complement, not replace, other heating systems. |

## Now test yourself

TESTED ☐

20 A website article claims that solar power supplies 'free electricity'. Explain why this statement is not true.
21 Students suggest building a wind turbine on school grounds. What disadvantages might they need to consider?
22 Why is nuclear power considered a non-renewable resource?

Answers on page 135

## Summary

- Energy is transferred between stores (chemical, kinetic, gravitational, elastic, thermal, magnetic, electrostatic, nuclear) by physical processes (including mechanical working, electrical working, heating by particles, radiation and waves).
- Energy is measured in joules (J).
- The principle of conservation of energy states that energy cannot be created or destroyed, only transferred.

- $$\text{efficiency} = \frac{\text{useful energy output}}{\text{total energy input}}$$

- Conduction and convection involve heating by particles. Heating may also occur by infrared (IR) radiation.
- Metals have free electrons that make them good conductors. Gases, like air, are poor conductors and may be described as insulators.
- Liquids and gases expand when heated and these heated regions rise. The patterns of movement are called convection currents.

- Black and matt (dull) surfaces are good emitters and absorbers of thermal radiation. White or shiny surfaces are poor emitters and absorbers, and are sometimes called reflectors.
- work = force × distance
- Work done, $W$, is measured in joules (J). One newton exerted to move one metre does one joule of work.
- For a gravitational store,
  $\text{GPE} = m \times g \times h$
- For a kinetic store,
  $\text{KE} = \frac{1}{2} \times m \times v^2$
- $$\text{power} = \frac{\text{work done}}{\text{time}}$$
- Power, $P$, is measured in watts (W). One watt means one joule is transferred each second.
- Electricity can be generated by a range of methods, some renewable and some non-renewable. Some are better for generating electricity on a large scale and others are better for single buildings. All methods have advantages and disadvantages to consider including cost, reliability and environmental impact.

## Exam practice

1 A local council suggests replacing a coal-fired power station with a 'solar farm'.

(a) (i) The planned solar farm would supply an average of 5 MW over a year. Explain why this figure is misleading. [1]

(ii) The average output of the coal-fired power station is 2000 MW. Calculate how many joules of energy are transferred in a minute. [3]

(b) Some campaigners claim that the solar farm would be expensive, others that it would be almost free. Explain how each claim could be justified. [2]

(c) The council decides to build the solar farm without shutting down the coal-fired power station. Explain how this will benefit the local environment. [2]

2 A model car is launched up a ramp using an elastic band. The velocity of the car is measured as it is released at the bottom of the ramp.
(a) Describe how energy is transferred to the moving car. [2]
(b) The initial velocity is measured as 4 m/s and the mass of the car is 1.2 kg. Calculate the kinetic energy. [4]
(c) Ignoring air resistance and friction, calculate what the maximum increase in height will be for the model car. [4]

3 All light bulbs cause heating of the room. What matters is how that energy is transferred.
(a) Describe how energy is usefully transferred by a light bulb in a room. [1]
(b) (i) A bulb with a power rating of 9 W is left on for 5 minutes. Calculate how much energy is transferred in this time. [3]
(ii) If the efficiency is 25%, calculate how much energy is wasted in this time. [3]
(iii) A new design transfers the same useful amount but is 50% efficient. Calculate the total energy supplied each second. [4]

4 When a battery-powered motor is used to lift a load, some of the energy heats the components and the surroundings.
(a) What word is used to describe this effect? [1]
(b) Identify the stores and processes labelled in this diagram. [3]

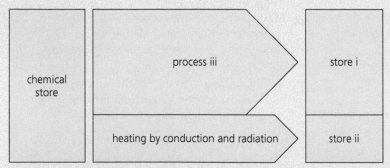

(c) (i) During use, 500 J is transferred from the chemical store and a total of 150 J is transferred by conduction and radiation. Calculate how much has been transferred to store (i). [2]
(ii) Calculate the efficiency of the motor. [3]

5 Two friends are pushing a broken down car.
(a) Together they exert a force of 600 N to move the car 20 m. Calculate how much work they have done. [3]
(b) Identify which force they are overcoming during this time. [1]
(c) Calculate their power (in watts) if this takes 30 seconds. [3]

## Answers and quick quizzes online

ONLINE

# 5 Solids, liquids and gases

## Density

### What is density?

If two identically sized samples of iron and wood are compared, one will have a much greater mass. The iron is said to have a greater **density**. This property is often what we mean when a material, rather than an object, is described as 'heavy'.

Density can be calculated by measuring the mass of an object and the volume it takes up.

$$\text{density} = \frac{\text{mass}}{\text{volume}}$$

$$\rho = \frac{m}{V}$$

> **Density:** The density of a material or object is calculated by dividing the mass in kilograms by the volume in metres cubed. The symbol $\rho$ is used and the standard unit of density is kilograms per metre cubed (kg/m³).

density, $\rho$, measured in kilograms per metre cubed (kg/m³)

mass, $m$, measured in kilograms (kg)

volume, $V$, measured in metres cubed (m³)

#### Example

A cube of aluminium, each side 4 cm, has a mass of 160 g. What is the density?

##### Answer

$$\rho = \frac{m}{V}$$

$$\rho = \frac{0.16}{0.04 \times 0.04 \times 0.04}$$

$$\rho = 2500 \, \text{kg/m}^3$$

> **Exam tip**
>
> When measuring small samples, it may be easier to find the mass in grams (g) and the volume in centimetres cubed (cm³). These measurements must be converted into standard units or the equation will then give density in non-standard units, grams per centimetre cubed (g/cm³). Alternatively, a non-standard answer can also be converted to standard units: 1000 kg/m³ = 1 g/cm³.

## Required practical

### Investigate density using direct measurements of mass 1: solids

#### Method

1 The mass of a block of wood was measured using an electronic scale.
2 The dimensions of the block were measured carefully with an accurate ruler.

#### Results

The mass was found to be 173.2 g or 0.1732 kg. The dimensions of the block were as shown in the diagram.

The density of the block was calculated as 2750 kg/m³ (to 3 s.f.), slightly greater than the value for aluminium in the previous Example. This shows that wood is not always less dense than metal.

**Figure 5.1 The dimensions of the block**

Making strong objects is relatively easy. The problem is that strong often means dense, which can restrict the use of strong materials. Some designed materials such as aluminium alloys and fibreglass are less dense than equivalents such as steel. These materials can be used for vehicles such as aircraft and racing yachts, but are usually too expensive for large engineering projects.

If the density of an irregular solid is required, the volume can be found by measuring how much liquid it displaces when added to a measuring cylinder or beaker.

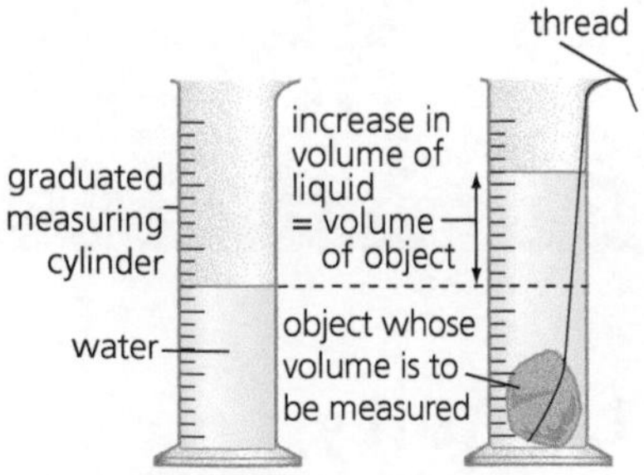

**Figure 5.2 Measuring the volume of a small irregularly shaped object**

## Required practical

### Investigate density using direct measurements of mass 2: liquids

#### Method

1 An empty 100 ml measuring cylinder was placed on an electronic balance. The mass was recorded.
2 Ethanol was added to the measuring cylinder up to the 20 ml line. The total mass (ethanol and cylinder) was recorded.
3 The mass of the liquid was calculated by comparing the first two measurements.
4 More ethanol was added to the 40 ml line. The total mass was recorded and the mass of the ethanol was calculated. This was repeated for 60, 80 and 100 ml.

#### Results

| Volume (ml) | Mass of measuring cylinder + liquid (g) | Mass of liquid (g) | Density (g/cm$^3$) |
|---|---|---|---|
| 0 | 34.2 | | |
| 20 | 49.8 | 15.6 | 0.780 |
| 40 | 65.3 | 31.1 | 0.778 |
| 60 | 81.7 | 47.5 | 0.792 |
| 80 | 97.4 | 63.2 | 0.790 |
| 100 | 113.0 | 78.8 | 0.788 |

The average density works out at 0.786 g/cm$^3$ or 786 kg/m$^3$. This is very close to the accepted value for the density of ethanol of 789 kg/m$^3$.

> **Exam tip**
>
> Remember that 1 centimetre cubed (cm$^3$) is the same as 1 millilitre (ml). Pure water at room temperature has a density of 1 g/cm$^3$ or 1000 kg/m$^3$.

## Now test yourself

TESTED

1 A student compares two equal boxes, one filled with feathers and the other with gravel. Explain the mass difference between the two, using the idea of density.
2 A sample cuboid, size 6 cm × 10 cm × 6 cm, is made of steel with a density of 8000 kg/m$^3$. Calculate the mass.
3 In food testing, a 100 ml sample is made up of a mixture of ethanol (density 0.789 g/cm$^3$) and water (which has a density of 1 g/cm$^3$). If 40 ml of the mixture is ethanol, what will the total mass be?

Answers on pages 135–6

# Pressure

## Pressure points

REVISED

When you push a drawing pin into a cork board, the same force acts at the point as on your thumb. The reason the forces have a different effect is that the area they are exerted over is different. The smaller the area, the greater the **pressure** caused by a force.

$$\text{pressure} = \frac{\text{force}}{\text{area}}$$

$$P = \frac{F}{A}$$

pressure, $P$, measured in newtons per metre squared ($N/m^2$)

force, $F$, measured in newtons ($N$)

area, $A$, measured in metres squared ($m^2$)

> **Pressure**: Describes how a force is concentrated into a contact area. Pressure is measured in newtons per metre squared ($N/m^2$) which is the same as the pascal (Pa).

### Example

A box weighing 35 N has a surface area of $7\,cm^2$ in contact with the table. How much pressure is acting?

Answer

$$P = \frac{F}{A}$$

$$P = \frac{35}{0.0007}$$

$$P = 50\,000\,N/m^2 \text{ or } 50\,000\,Pa \text{ or } 50\,kPa$$

> **Exam tip**
>
> Remember that 1 metre is 100 cm. This means that a metre squared is 100 cm × 100 cm or $10\,000\,cm^2$.

### Typical mistake

There may be confusion between the units of pressure (newtons per metre squared, $N/m^2$) and those for the stiffness of a spring (newtons per metre, N/m) and turning moment (newton metres, Nm). Using **pascals** (Pa) helps to avoid the confusion. If unsure, check the units of the variables in the equation. Remember that force in newtons divided by area in metres squared gives pressure in newtons per metre squared.

> **Pascal**: The unit of pressure, equal to one newton of force applied to a contact area of one square metre. Most examples will use kilopascals (kPa), where 1 kPa = 1000 Pa.

### Now test yourself

TESTED

4 (a) Rearrange the equation for pressure so that force is the subject.
  (b) A tiled floor can withstand a maximum pressure of 400 kPa without being damaged. If four table legs have a total surface area of $12\,cm^2$, what is the maximum total weight, including the table itself, that should be supported by the legs?
5 Why are press-ups harder if you lean on your fingertips rather than on the palms of your hands?
6 Explain why both polar bears and camels have evolved to have very wide feet.

Answers on page 136

# Pressure in liquids and gases

If materials are rigid, a force on them acts in a linear direction only. If a material is a fluid (a liquid or a gas) then a force acting on the material will cause the pressure against every surface to increase. When a force causes pressure elsewhere, we say that the pressure has been transmitted. The pressure in a fluid, whatever the value, acts equally in all directions.

# Increase of pressure with depth

The pressure on an object increases as it is lowered beneath the surface of a liquid like water. The deeper it is below the surface, the greater the pressure. This is due to the weight of water above it. Mathematically, we can treat this as a vertical column of water. This force acts on all sides of the object at once. Different liquids cause different forces because of differences in their density.

$$P = h \times \rho \times g$$

pressure, $\rho$, measured in newtons per metre squared ($N/m^2$) or pascals (Pa)

height of the column, $h$, measured in metres (m)

density of the liquid, $\rho$, measured in kilograms per metre cubed ($kg/m^3$)

gravitational field strength, $g$, measured in newtons per kilogram (N/kg); on Earth $g = 10\,N/kg$

### Example

A watch falls to the bottom of a lake, 25 m below the surface. Water has a density of $1000\,kg/m^3$. What is the pressure on the watch due to the water?

Answer

$$P = h \times \rho \times g$$
$$P = 25 \times 1000 \times 10$$
$$P = 250\,000\,Pa \text{ or } 250\,kPa$$

This might seem like a large pressure, but in fact the watch was under pressure even before it fell in the water. At this depth the watch is actually experiencing a total pressure of 350 kPa: 100 kPa from the air and 250 kPa from the water.

## Measuring atmospheric pressure

Air is a mixture of gases and so it acts as a fluid. This means there is a force acting on objects, which is the weight of the atmosphere. The density of air is much lower than that of a liquid like water, but the height of the vertical column of air is large.

At sea level, the pressure created by the atmosphere is around 100 kPa. This seems like a large value, but as it acts on every surface equally we tend not to notice it. If pressure changes very quickly, for example if you are sitting in a car that drives down a hill at high speed, you may notice the effect when your ears 'pop'. In the mountains, the column of air above us is not as high and so air pressure is lower.

## Hydraulic machines

The shape of a liquid changes but the volume of the liquid is almost constant, no matter how much it is compressed. If the liquid is sealed in a container, a force on one surface will cause pressure to act equally in all directions, even around corners. This effect is used to change the direction and magnitude of a force in a hydraulic system, for example to control brakes in a car.

A hydraulic system uses tubes or pipes with different diameters, so the area is different for the input and output forces. Pushing down on the liquid surface in a small pipe causes a pressure to act on the liquid. The same pressure then acts on all the other surfaces, including the larger pipe, so an upwards force is caused that is greater than the input.

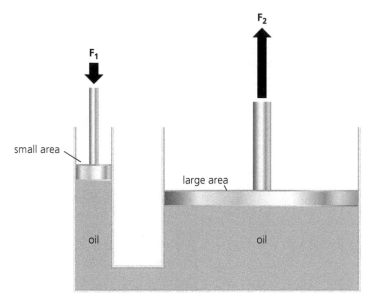

**Figure 5.3** A small force acting on a small area causes a large force to act because of the larger area.

**Exam tip**

It can be easy to miss, but remember that the input piston moves much further than the output piston. This must be true because the work done by the first piston must be the same as the work done by the second one, otherwise the principle of conservation of energy would be violated.

## Now test yourself

TESTED

7 A bottle with three holes in the side is filled with water. Why does the water travel further from the hole closest to the base as it leaks out?
8 If atmospheric pressure is around 100 kPa, why are we not pushed around by the force exerted by the air?
9 (a) Sea water has a density of 1025 kg/m$^3$. How far under the sea must a diver swim before the water pressure becomes 100 kPa?
   (b) Why is the total pressure on the diver at this distance actually 200 kPa?

Answers on page 136

# Solid, liquid and gaseous states

Substances can exist in three different **states**. The chemical properties of a substance don't change, nor do the individual particles. The particles stay the same, but their arrangement and movement changes.

- In a solid, the particles are close together and sit in a regular pattern. The particles are held in place by attractive forces and cannot move around, but can vibrate. It is difficult to change the shape of a solid.
- In a liquid, the particles are still close together, but the forces are weaker so the particles can move around. The liquid changes shape depending on the container, and will flow when poured. Liquids are very hard to compress.
- In a gas, the particles are moving much faster and, although they often collide with each other, the gaps between them are on average much larger. A gas expands to fill any container but can also be compressed.

**State:** The physical arrangement and motion of particles in a substance. Solid, liquid and gas are considered the normal three states of matter.

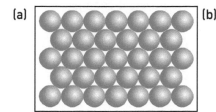 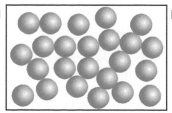 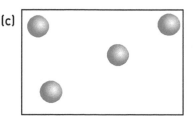

**Figure 5.4** The particle arrangement in (a) a solid, (b) a liquid and (c) a gas

Solids and liquids are denser than gases because the gaps between particles are small, which means there are more particles in the same volume. When comparing two substances in the same state, it is the mass of each particle that matters: gold atoms have more mass than aluminium, so gold is denser than aluminium.

## Internal energy

The particles in a system can be an energy store. (See page 61 for a reminder about energy stores.) This is called **internal energy** and can be thought of as a combination of the energy in the kinetic store (because the particles are moving) and the potential store (because of the attractive forces between the particles). The internal energy changes when a material is heated or cooled because energy is transferred between the material and surroundings.

When a material is heated, the internal energy increases. Sometimes this is measured as an increase in the temperature of the system, because the particles are moving (or vibrating) faster. Sometimes heating causes the particles to move further apart, and this is observed as a change of state.

> **Internal energy**: The total energy in the kinetic and potential stores of a material. It is increased when a material is heated, whether it changes the temperature or the state.

### Changes of state

Each change of state is linked to a change in the internal energy of the substance that affects the separation of particles and how quickly they move.

When substances are heated, solids **melt** to form liquids and liquids **boil** or **evaporate** to form gases. Some substances turn straight from a solid to a gas without a liquid state, which is called **sublimation**, but this is a rare occurrence.

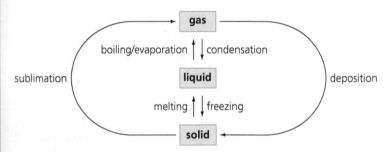

**Figure 5.5 Each arrow is labelled with a change of state.**

When substances are cooled the internal energy decreases. Gases **condense** to form liquids and liquids **freeze** to form solids. Where substances turn straight from a gas to a solid without a liquid state, this is called **deposition**.

These changes in state are all **physical changes** – no new substance is produced and the process is reversible. Mass is always conserved, there are no chemical changes, and the arrangements of atoms and molecules do not alter. However, there is an internal energy change, increasing from solid to liquid and increasing greatly from liquid to gas.

> **Melt**: Change of state from a solid to a liquid
>
> **Boil** or **evaporate**: Change of state from a liquid to a gas
>
> **Sublimation**: Change of state from a solid directly to a gas. This only happens in a few examples, such as carbon dioxide.
>
> **Condense**: Change of state from a gas to a liquid
>
> **Freeze**: Change of state from a liquid to a solid
>
> **Deposition**: Change of state from a gas directly to a solid
>
> **Physical change**: A change of state that is reversible, unlike a chemical reaction, and no new substances are formed. Heating and cooling are physical changes, as are changes of state.

> **Typical mistake**
>
> Students often lose marks because they describe a temperature change (heating or cooling) rather than being specific about a change of state. When solid ice in a warm room melts and becomes liquid water, the change of state is important. Saying that the ice has warmed up is unlikely to gain all the available marks.

## Required practical

# Obtain a temperature–time graph to show the constant temperature during a change of state

### Method

A hot object in a cool room transfers energy to the surroundings. In this experiment, the hot sample does not just cool but also changes state from liquid to solid.

1 All students put on safety glasses.
2 The tripod and gauze were set up and a beaker with 150 ml of water was heated.
3 A second tripod and gauze was set up to hold a beaker with ice water.
4 A test tube of stearic acid, covered with a cotton wool plug, was clamped in a retort stand and placed in the hot water.
5 Once the stearic acid reached 100 °C, as confirmed with a thermometer, the test tube was placed in the ice water and the stopwatch was started. At this temperature, the stearic acid was a liquid.
6 The temperature of the stearic acid was recorded every minute until it reached 50 °C.
7 A graph was drawn with time on the horizontal axis and temperature on the vertical axis.

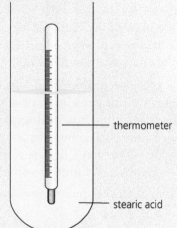

thermometer

stearic acid

**Figure 5.6 The temperature change will happen faster if there is ice around the boiling tube.**

### Results

The stearic acid cooled quickly at first (shown by a steep line down on the graph). For several minutes the temperature then remained constant (shown by a flat line on the graph), and during this time the stearic acid turned from a liquid to a solid. The stearic acid then cooled further, but not as quickly as before (shown by a shallow line down).

During the cooling process, internal energy was being transferred to the surroundings. The part of the graph that was flat, with a constant temperature, shows where the stearic acid was changing state rather than cooling down. The energy was transferred from the potential store of the particles. Changes of state for a pure substance always happen at a fixed temperature.

## Now test yourself

TESTED

10 What word describes a state change from
  (a) liquid to solid
  (b) gas to liquid
  (c) gas to solid?
11 Ice from a freezer at a temperature of –15 °C is placed in a warm room and the temperature is recorded over several hours. At which temperature would the first flat line be and why?
12 Particles in all substances could be described as 'moving', but they are only moving around in liquids and gases. Give the word used to describe how the particles move in a solid and explain what it means.
13 What happens to the gaps between particles when steam (a gas) condenses on a cold mirror to form water droplets (a liquid)?

Answers on page 136

### Exam tip

If a question includes a temperature graph with a state change, you may find it helpful to add notes about where the substance is warming or cooling and where it is changing state. This will help you to write a clear description in a logical sequence.

# Specific heat capacity

When a material is heated, the internal energy increases. When it cools, the internal energy decreases again as energy is transferred to the surroundings. If there is no state change, the change occurs in the thermal store and depends on:

- the mass of the material
- what the substance is made of
- the change in temperature.

The energy required to change the temperature of a material by one degree Celsius per kilogram of mass is the **specific heat capacity**. This has a unit of joules per kilogram degrees Celsius (J/kg°C). Materials with a low specific heat capacity, such as copper, heat up and cool down quickly. Water has a high specific heat capacity and requires lots of energy to be transferred for even a small change in temperature.

change in thermal energy = mass × specific heat capacity
× temperature change

$$\Delta Q = m \times c \times \Delta T$$

## Example

A concrete block, mass 70 kg, is heated from 2 °C in the morning to 14 °C in the early afternoon. Concrete has a specific heat capacity of 800 J/kg °C. How much energy has been transferred in this time?

### Answer

$\Delta Q = m \times c \times \Delta T$

$\Delta Q = 70 \times 800 \times (14 - 2)$

$\Delta Q = 70 \times 800 \times 12$

$\Delta Q = 672\,000\,\text{J}$ or 672 kJ

This is an example of an equation that describes the energy transferred to or from an energy store. See page 68 for more information.

## Required practical

### Investigate the specific heat capacity of materials, including water and some solids

There are many ways to heat a material, but to do so in a measurable way means using an electric heater with a voltmeter and an ammeter included in the circuit. See page 25 for more information about how to calculate the energy transferred by an electric current.

### Method

1. 100 ml (0.1 kg) of water was poured into a polystyrene cup with a lid.
2. The electrical heater was placed into the water, with an ammeter placed in series and a voltmeter placed in parallel.
3. The water was stirred and the temperature, $T_1$, recorded.
4. The heater was switched on for 5 minutes, then turned off. The voltmeter and ammeter readings were recorded while the heater was switched on.
5. The water was stirred and once the maximum temperature was reached it was recorded as $T_2$. ⇨

---

**Revision activity**

You could use a flow chart or annotated sketch for summary notes; the point is to focus on the important points so you can expand on them under pressure. Test yourself by completing a partially completed diagram. As you become more confident, start from a version with fewer hints.

---

**Specific heat capacity, $c$:** The energy needed to change the temperature of one kilogram of a material by one degree Celsius. Measured in joules per kilogram degrees Celsius (J/kg°C).

---

change in thermal energy, $\Delta Q$, measured in joules (J)

mass, $m$, measured in kilograms (kg)

specific heat capacity, $c$, measured in joules per kilogram degrees Celsius (J/kg°C)

change in temperature, $\Delta T$, measured in degrees Celsius (°C)

---

**Typical mistake**

Students often use a single temperature reading in calculations, instead of comparing two values to find the temperature change. When you revise, remind yourself that $\Delta T$ means **a change in** temperature.

**6** This method was repeated for solid blocks of sample metals, which were wrapped in insulation to reduce thermal transfer to the surroundings.

## Results

The temperature change of the water was calculated. The starting temperature was 14 °C and the final temperature was 23 °C.

$$\Delta T = T_2 - T_1$$
$$\Delta T = 23 - 14$$
$$\Delta T = 9\,°C$$

The energy transferred by the heater was calculated. The readings show that a current of 1.3 A flowed for 5 minutes and there was 11.8 V across the heater.

$$E = I \times V \times t$$
$$E = 1.3 \times 11.8 \times (5 \times 60)$$
$$E = 1.3 \times 11.8 \times 300$$
$$E = 4602\,J$$

These values were used to calculate the specific heat capacity of water, based on the energy transferred electrically, $E$, being the same as the change in the thermal store, $\Delta Q$.

$$\Delta Q = m \times c \times \Delta T$$
$$c = \frac{\Delta Q}{m \times \Delta T}$$
$$c = \frac{4602}{0.1 \times 9}$$
$$c = 5113\,J/kg\,°C \text{ or } 5100\,J/kg\,°C \text{ (to 2 s.f.)}$$

This is higher than the accepted value of 4200 J/kg °C. This difference can be explained since energy is transferred to the environment from the water, which reduces the final temperature, $T_2$. The readings on the meter may not have been exact, and d.c. meters usually give a lower value when an alternating current is being measured.

The readings taken during the heating of metals also gave higher values for the heat capacity, when compared to the accepted values.

> **Exam tip**
>
> Always leave rounding until your final answer. Get into the habit of giving your answers to the same level of accuracy, in this case two significant figures, as the data provided in the question.

## Now test yourself

TESTED

14 Equal masses of water and copper are heated by identical Bunsen burners for the same amount of time. Which one will have a higher final temperature?

15 If the sample from Question 14 contained 150 ml of water and the temperature rose by 40 °C, how much energy was transferred? For water, $c = 4200\,J/kg\,°C$.

16 A student completed the required practical described on page 82, but used a beaker rather than an insulated cup. What difference would this make to:
(a) the measured temperature change
(b) the calculated specific heat capacity?

Answers on page 136

# Ideal gas molecules

## The particle model of gases

REVISED

Scientific models allow us to explain what happens and make testable predictions. Models are never perfect, but they are useful simplifications for better understanding.

The particle model for gases, also called the kinetic theory (because it is about movement), helps to link the observed behaviour of gases to the particles that make them up. It does not matter whether the particles in question are atoms or molecules. In the model:

- Particles in a gas are constantly moving in random directions.
- The particles collide with each other and the walls of any container without transferring kinetic energy.
- The motion of the particles is related to the temperature; the higher the average kinetic energy of the particles, the higher the measured temperature of the gas.

Although the particles are very small, they still exert a tiny force on the walls of their container during each collision. There is no energy transferred to the walls because they do not move, so no work is done (see page 67 for more information). There are so many collisions that the total force on the area of the container is enough to measure, and this is detected as pressure.

> **Absolute zero**: The lowest possible temperature, equal to –273 °C, and the starting point for the kelvin temperature scale

## Linking the pressure of a gas to its temperature

REVISED

If a sample of gas in a sealed container is heated or cooled, the pressure it exerts on the walls of the container changes. (If the container is flexible then the volume will change instead or as well.) By keeping the volume of the gas the same, and by recording the pressure for different temperatures, a graph can be plotted. The points should make a straight line. According to the model, this shows that particles at a higher temperature move faster because they have more energy. More energy means more collisions, which means a higher pressure on the container walls.

The straight-line graph predicts that at a very low temperature, –273 °C, the pressure will be zero. The only way to explain this is that the particles stop moving completely. This temperature is called **absolute zero** and is the lowest possible temperature of anything.

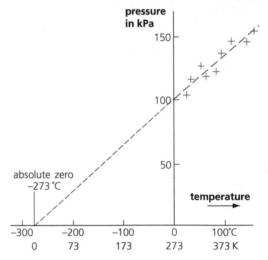

**Figure 5.7 As temperature of a fixed volume of gas increases, pressure increases.**

## The Kelvin scale

REVISED

Temperature is measured from the lowest possible value, absolute zero. When the particles have stopped moving they have zero kinetic energy. Physicists call this the **absolute temperature scale**, and it is useful because it means we can see how gas pressure is directly proportional to absolute temperature. Absolute zero is 0 **kelvin**, 0 K, which means 0 °C is 273 K.

> **Absolute temperature scale**: A scale used to measure temperature with units kelvin (K)
>
> **Kelvin (K)**: The unit of absolute temperature. The size of a kelvin is the same as 1 °C.

### Exam tip

When writing temperatures in kelvin, do not use a degree sign. The units are kelvins (K) not 'degrees kelvin'.

temperature in kelvin = temperature in degree Celsius + 273

# Calculating the pressure

When the volume is kept constant there is a relationship between the temperature and pressure of a gas. This relationship is often called Charles' law. The ratio of pressure to temperature stays the same as the gas is heated or cooled. The formula only works if the absolute temperature scale is used.

$$\frac{P_1}{T_1} = \frac{P_2}{T_2}$$

> pressure, $P$, measured in pascals (Pa)
>
> temperature, $T$, measured in kelvin (K)
>
> The subscripts 1 and 2 are used to show the values 'before' and 'after' a change.

### Example

At 300 K, a gas sample in a canister exerts a pressure of 90 kPa on the inside surface. What will the new pressure be if it is heated to 350 K?

Answer

$$\frac{P_1}{T_1} = \frac{P_2}{T_2}$$

$$P_2 = T_2 \times \frac{P_1}{T_1}$$

$$P_2 = 350 \times \frac{90}{300}$$

$$P_2 = 105 \, kPa$$

> **Exam tip**
>
> If you use kilopascals for one pressure value then the calculated value for the other will also be in kilopascals.

# Linking the volume of a gas to its pressure

If a sample of gas is kept at the same temperature, the relationship between pressure and volume can be investigated. As the volume is decreased, the pressure increases. At a particle level, the smaller space means collisions happen more often, and so the total force increases.

This relationship is often called Boyle's law. For any sample that is kept at constant temperature, multiplying the pressure by the volume will give the same value as it is compressed.

$$P_1 \times V_1 = P_2 \times V_2$$

> pressure, $P$, measured in pascals (Pa)
>
> volume, $V$, measured in metres cubed ($m^3$)

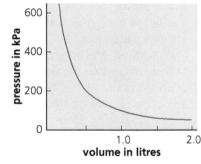

**Figure 5.8** When temperature is constant, halving the volume of the gas doubles the pressure.

### Example

A sample of gas at 150 kPa takes up 0.67 litres. The volume is decreased to 0.25 litres while the temperature is kept the same. What is the new pressure?

Answer

$$P_1 \times V_1 = P_2 \times V_2$$

$$P_2 = \frac{P_1 \times V_1}{V_2}$$

$$P_2 = \frac{150 \times 0.67}{0.25}$$

$$P_2 = 400 \, kPa$$

> **Exam tip**
>
> As before, different units can be used as long you remember that the answer will also be in those units.

## Now test yourself

TESTED

17 Why does a car tyre that was fully inflated in the winter have a higher chance of bursting in the summer?
18 Nitrogen gas condenses to a liquid at –196 °C.
   (a) What is this value on the absolute temperature scale?
   (b) Give the melting and boiling points of water in kelvin.
19 A gas canister has a faulty valve and half of the gas leaks into the air. Why does the canister now feel cold? Use the movement of particles to explain this change.
20 Which measured variable for a gas is directly proportional to temperature?

Answers on page 136

### Revision activity

Revise the relationships between pressure, temperature and volume by thinking of a simple example, such as a diver's air tank. Practise linking the observed changes to how the particles are behaving, using key words including speed, kinetic energy, force and area.

## Summary

- Density, $\rho$, is the mass of each unit volume of a material and has units of kilograms per metre cubed (kg/m$^3$).

$$\text{density} = \frac{\text{mass}}{\text{volume}}$$

- Pressure, $P$, is the force exerted per unit area of a surface and has units of newtons per metre squared (N/m$^2$) or pascals (Pa): 1 Pa = 1 N/m$^2$.
- The pressure caused by a column of a fluid (liquid or gas) depends on height, density and the force of gravity. Fluids exert pressure on all sides and surfaces equally.

$$P = h \times \rho \times g$$

- Materials can exist in different states and change from one state to another in a reversible physical change because of heating or cooling. Melting, evaporating, condensing, freezing, subliming and deposition are all changes of state.
- Particles change their position, separation and motion during state changes. These changes explain the properties of materials.
- Specific heat capacity, $c$, is the energy needed to raise the temperature of one kilogram of

a material by one degree Celsius. It has the unit of joules per kilogram degrees Celsius (J/kg °C).

$$\Delta Q = m \times c \times \Delta T$$

- Particles in a gas cause pressure because of many collisions with the surface of the container. The temperature, pressure and volume are all related.
- The absolute temperature scale starts from the lowest possible temperature, –273 °C, Known as absolute zero. At this temperature particles stop moving entirely. The units of this scale, kelvins (K), are the same size as degrees Celsius.
- If the volume is kept constant, increasing the temperature causes an increase in pressure. The volume is directly proportional to the absolute temperature.

$$\frac{P_1}{T_1} = \frac{P_2}{T_2}$$

- If the temperature is kept constant, increasing the pressure reduces the volume.

$$P_1 \times V_1 = P_2 \times V_2$$

## Exam practice

1 In a factory a steel component must be heated then cooled quickly. Steel has a specific heat capacity, $c$, of 450 J/kg °C.
   (a) Define specific heat capacity. [1]
   (b) Calculate how much energy is needed to heat the steel component, which has a mass 2.4 kg, from 20 °C to 650 °C. [3]
   (c) The hot steel glows red. State how the energy is being transferred. [1]
   (d) The component is immediately cooled down to 20 °C by placing it in 50 litres of cold water. Calculate the temperature change of the water. Assume that for water, $c$ = 4200 J/kg °C. [3]

2 In a cosmetics workshop a technician makes shapes with cocoa butter. It melts at 36 °C.
   (a) Describe the behaviour of the particles in the cocoa butter at 10 °C. [3]
   (b) A 750 ml beaker of the cocoa butter is placed in a water bath at 70 °C. Sketch the temperature of the sample over time. [3]
   (c) The density of solid cocoa butter is 0.96 g/cm³. Calculate the mass in the beaker. [3]
   (d) (i) Describe one change in the behaviour of particles in the liquid compared to the solid. [1]
       (ii) The liquid produced is poured into moulds and allowed to cool. What can you say about the mass of the solid at the end compared to the solid at the start? [1]

3 Engineers are designing a remote-controlled submersible to explore under the sea. It needs to be able to withstand the pressure of the water.
   (a) Which is the correct formula for pressure? [1]

   A $\quad P = \dfrac{F}{V}$

   B $\quad P = \dfrac{m}{A}$

   C $\quad P = \dfrac{F}{A}$

   D $\quad P = m \times V$

   (b) (i) An engineering apprentice suggests that only the top surface of the submersible needs to be strong, because the water is above it. Explain why they are wrong. [2]
       (ii) Calculate what the pressure from the water will be at 0.8 km below the surface. Use 1040 kg/m³ as the average density of sea water. [3]
   (c) A fish gets caught on the submersible and is pulled to the surface. Explain why it explodes. [2]

4 A medical supply company produces oxygen cylinders for patients in hospitals.
   (a) 460 litres of oxygen at a pressure of 100 kPa is supplied. Calculate the force in newtons that this would exert on a surface with area 1 m². [1]
   (b) The oxygen is kept at the same temperature and compressed to take up 2 litres in the cylinder. Calculate the new pressure. [3]
   (c) Explain why it is important that the filled cylinder is not left in direct sunlight. [2]
   (d) (i) A patient needs 4 litres of oxygen at 100 kPa per minute. Calculate whether the cylinder will last 2 hours. [3]
       (ii) Explain what will happen to the pressure inside the cylinder as the oxygen is used. [1]

5 A student is checking the density of different golf balls.
   (a) The student does not have the right equipment to measure the diameter of the golf balls. Describe a method, including any calculations, they could use to find the volume directly. [4]
   (b) State which equation they need to use to find the density. [1]
   (c) One make of golf ball is claimed to float in water. Explain what this suggests about the density of the ball. [1]

## Answers and quick quizzes online

ONLINE

# 6 Magnetism and electromagnetism

## Magnets

A magnet is an object that is attracted or repelled by other magnets, and will attract some materials. If a magnet is allowed to move freely, one end will point towards the Earth's North Pole (in the Arctic). This end is called the magnet's **north-seeking pole**. The other end will point in the opposite direction, towards the Earth's South Pole (in Antarctica). This end is called the **south-seeking pole**.

If a magnet is broken in half, each part will have its own north-seeking pole and south-seeking pole.

> **North-seeking pole**: The end of a magnet that is attracted to the Earth's North Pole, usually colour-coded red
>
> **South-seeking pole**: The end of a magnet that is attracted to the Earth's South Pole, usually colour-coded blue or white

## Poles

REVISED

If two magnets are placed close together, there are two possible effects. The poles can be attracted (pulled together) or repelled (pushed apart).

Two north-seeking poles will repel each other and so will two south-seeking poles. A north-seeking and a south-seeking pole are attracted to each other. This is often described as 'like poles repel, opposite poles attract'.

Magnets attract some materials that are not magnets themselves. These are called **magnetic** materials, and include iron, cobalt and nickel. This attraction is greatest when the material is close to the pole of a magnet. The two poles are equally strong and magnetic materials are always attracted to, not repelled by, a magnet.

Iron filings are attracted to the poles of a magnet. These filings can be used to see the **orientation** of the forces at different points around a magnet. To see the **direction** of the fields, a plotting compass is used instead. The region where magnets and magnetic materials experience a force is called the magnetic field.

> **Typical mistake**
>
> Magnets follow the same rules about attraction and repulsion as electric charges (see page 37 for more information) but magnetism and electrostatic charge are not the same. They are both examples of non-contact forces (see page 5 for more information) but they are caused by different fundamental properties.

> **Typical mistake**
>
> Make sure your explanation doesn't suggest that the iron filings create the magnetic field. All they do is show us where it is.

> **Magnetic**: Materials that are attracted to magnets, but are not themselves magnets. Iron, nickel and cobalt are magnetic, as are some rare earth elements like neodymium.

### Required practical

#### Investigate the magnetic field pattern for a permanent bar magnet and between two bar magnets

**Method**

1 A sheet of paper was placed over a bar magnet.
2 Iron filings were sprinkled over the paper. Every few moments the paper was tapped lightly.

3 The patterns formed by the iron filings were sketched, with particular attention paid to the areas around the poles.

4 The method was then repeated with two magnets in both possible arrangements with two similar poles together (two north-seeking poles or two south-seeking poles) and two opposite poles together (north-seeking and south-seeking).

## Results

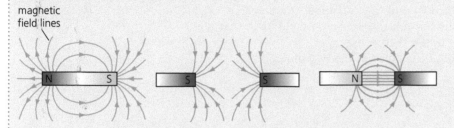

**Figure 6.1** The lines always start and end at a magnetic pole.

The magnetic field was strongest where the lines were closest together, at the poles. This is where the force on another magnet or a magnetic material is largest. The magnetic lines never cross, and they always connect a north-seeking pole to the closest south-seeking pole.

## Magnetic fields

REVISED

Some magnetic fields are stronger than others. Magnetic field lines are used to represent the strength and direction of the field. Like the arrow that might be drawn to show weight acting towards the ground because of gravity, they are a model rather than suggesting magnetism only works along those specific lines.

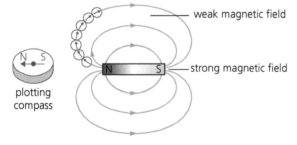

**Figure 6.2** Plotting field lines around a bar magnet

Plotting compasses are used in drawings to show the direction of the magnetic field lines. Arrows should be added to point from a north-seeking pole to a south-seeking pole.

If two magnets are close together the fields will affect each other. For example, if similar poles are close there will be a neutral area in between them (X in Figure 6.3) where there is no magnetic field. If opposite poles are close the field between them will be uniform, which means that the field has the same strength in this area and the field goes in one direction.

> **Typical mistake**
>
> Don't use the term 'big magnet' when you mean 'strong magnet'. Although these terms might seem similar in meaning, writing about the strength of a magnet shows you understand it is the effect of the field which matters.

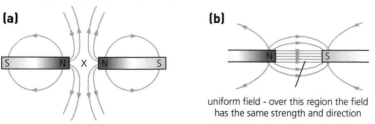

**Figure 6.3** (a) A neutral point, X, between two similar poles; (b) a uniform field between two opposite poles

## Now test yourself

TESTED

1 Predict what will happen when the north-seeking pole of a magnet is used to test the following materials:
   (a) iron nail
   (b) wooden ruler
   (c) copper pipe
   (d) another magnet.
2 (a) Sketch the magnetic field lines around a bar magnet, ignoring the direction arrows.
   (b) How would the diagram be different for a stronger magnet?
3 Two magnets are positioned so there is a uniform field between them.
   (a) Sketch the field lines that this will create.
   (b) What does this tell you about the poles of the two magnets?

Answers on page 136

# Magnetising

## Magnetic domains

REVISED

A **magnetically soft** material can become an induced (or temporary) magnet when in a magnetic field. A piece of pure iron magnetised like this stops acting like a magnet relatively quickly. In contrast, steel is a **magnetically hard** material and, once magnetised, will attract or repel other magnets until it is demagnetised.

If a magnet is used to pick up some pins which are pure iron, they will become induced magnets as long as they are in contact with the magnet. As soon as the magnet is removed, the effect stops. If instead some steel nails are picked up, they will become weak but permanent magnets. This effect will last even without the original magnet nearby, although it can be removed by heating the metal or hitting it with a hammer. This is because the magnetic effect is caused by lined up regions within the material called domains. These domains become disrupted by heating or beating the magnet.

> **Magnetically soft:** Describes materials that can become temporary (or induced) magnets, for example iron
>
> **Magnetically hard:** Describes materials that can be magnetised to become permanent magnets, until demagnetised, for example steel

## Magnetising

REVISED

For centuries it was known that a piece of unmagnetised steel, if stroked in one direction with a magnet, could generate its own magnetic field. An easier way of creating a similar field involves putting a piece of steel in a coil of wire called a **solenoid**. A large current in the wire makes a temporary but strong magnetic field, which permanently magnetises the steel.

A solenoid is also part of an **electromagnet**, but this includes an iron bar instead of steel. Because iron is magnetically soft, it becomes a strong induced magnet *only* while there is current in the solenoid. Electromagnets are frequently used in industry where the ability to turn a magnet on and off helps complete tasks, for example when lifting and dropping materials in a recycling centre.

> **Solenoid:** A coil of wire that is part of an electric circuit. While current is flowing it has a magnetic field around it similar to a permanent bar magnet. The more turns or loops of wire in the coil, the bigger the effect.
>
> **Electromagnet:** A solenoid with an iron core, which has a magnetic field only when current is flowing

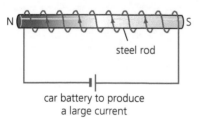

steel rod

car battery to produce a large current

**Figure 6.4 The piece of steel becomes a permanent magnet after the current has been switched on and off.**

**Now test yourself**

TESTED

4  Explain the difference between a permanent magnet and an induced magnet.
5  Why do the pins shown in Figure 6.5 repel each other?
6  Would you use a magnetically hard or soft material to make an electromagnet? Why?

Answers on page 136

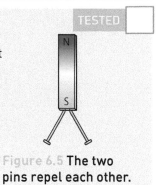

Figure 6.5 **The two pins repel each other.**

# Currents and magnetism

## Magnetic fields near a straight wire

REVISED

The magnetic field lines around a bar magnet go from the north-seeking pole to the south-seeking pole. A magnetic field is also caused around any wire which has current flowing, but the lines form concentric rings around the wire. The field is strongest close to the wire.

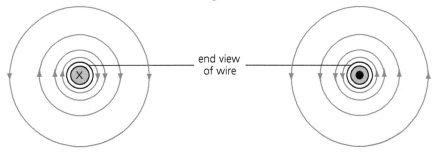

end view of wire

Figure 6.6 **The dot and cross show the direction of the current in the wire. The arrows show the direction of the magnetic field lines.**

The direction of the magnetic field can be worked out if we know the direction of the current. Figure 6.6 shows the wire with the current direction going into the paper (indicated by the 'X') and the field lines pointing clockwise around the wire. If the wire has the current flowing out of the paper (indicated by the 'dot') the field lines point anti-clockwise around the wire. This can be remembered using the **right-hand grip rule**, whereby your thumb points in the direction of current and your curled fingers point in the direction of the magnetic field lines.

> **Exam tip**
>
> When describing how an electromagnet works, it is important to include the current flowing in the coil. Students can lose marks if they imply that the iron core or the copper wires cause the magnetism.

> **Right-hand grip rule**: A way to remember the direction of magnetic field lines around a current-carrying wire

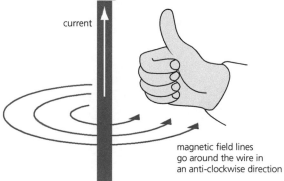

current

magnetic field lines go around the wire in an anti-clockwise direction

Figure 6.7 **The right-hand grip rule.**

## Magnetic fields near coils of wire

REVISED

A single loop in a coil of wire acts as if it has two magnetic fields, each making a circle around one side of the loop. The right-hand grip rule helps us work out the direction for each field. Because the current in the loop is continuous, the two fields add up and so there are magnetic lines created through the centre of the loop like you would see in a bar magnet. For a single loop, these magnetic lines will be very weak.

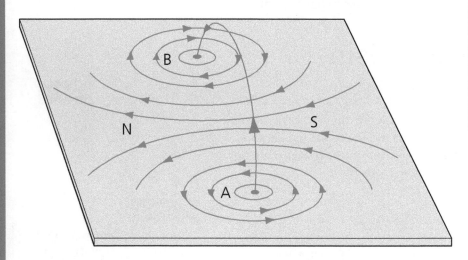

**Figure 6.8 The magnetic field lines near a single loop of wire (current flowing from A to B)**

If there are two or more loops of wire, they make a coil called a solenoid. As the coil gets longer, with more loops or turns, the magnetic field looks more and more like the one seen around a bar magnet.

## Producing large magnetic fields

REVISED

An electromagnet produces a stronger magnetic field if it has:
- a larger current
- more turns of wire
- iron in the middle of the solenoid.

Some of these changes are easier to make in practical terms than others. For example, a high current may cause the wire to melt or a fuse to blow. If very strong magnetic fields are needed, for example in particle accelerators, then wires are made of superconducting materials that have zero resistance. These materials must be kept very cold, either with liquid nitrogen (to cool the wires to 77 K) or liquid helium (4 K).

## Now test yourself

7  Where is a magnetic field strongest for a solenoid?
8  Explain the meaning of the three symbols that are used to show the direction of current in a wire: arrows on a line, circle with a cross and circle with a dot.

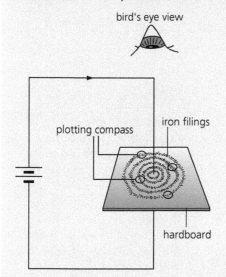

Figure 6.9 **The magnetic field lines around a straight wire make concentric rings.**

9  In Figure 6.9 which direction would the magnetic field lines point? Explain how this can be worked out without looking at the plotting compasses.
10 A student is trying to make an electromagnet stronger but cannot increase the current because the power supply fuse keeps blowing.
   (a) Suggest what else they could change.
   (b) How might they measure the strength of the electromagnet?

Answers on pages 136–7

# The motor effect

## Combining two magnetic fields

REVISED

We know that when a current flows in a wire, a magnetic field is produced. However, if there is another magnetic field present, the two fields will interact and cause a force. Unless both objects are fixed in place, this force will cause movement. This movement is called the **motor effect** and is the basis of almost every situation in which an electrical current causes motion, from battery-powered fans to loudspeakers, from washing machines to the propulsion of aircraft carriers.

> **Motor effect:** The use of two magnetic fields, one permanent and one from an electromagnet, to cause movement

The motor effect is easiest to observe with a single wire. There is a uniform field between the poles of the permanent magnet(s). The wire, and the current, is at right angles to the magnetic field. The movement of the wire is at right angles to both. If the current reverses direction, so does the movement of the wire.

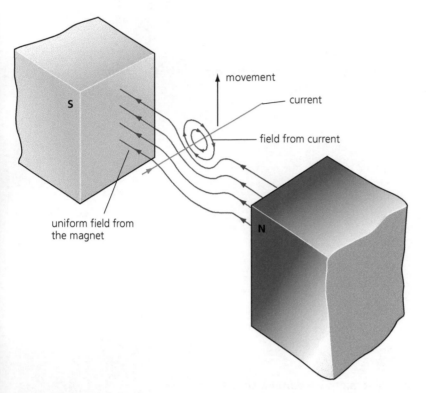

movement

current

field from current

uniform field from
the magnet

S

N

**Figure 6.10 The movement is at right angles to the directions of the magnetic field and the electrical current.**

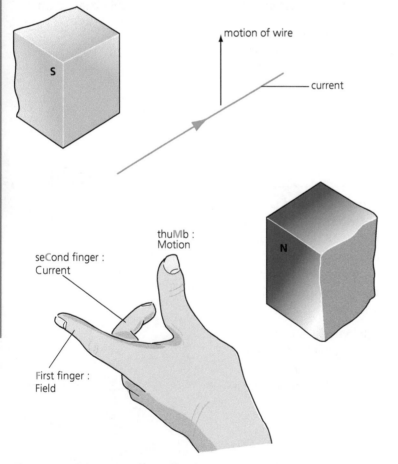

motion of wire

current

S

N

thuMb :
Motion

seCond finger :
Current

First finger :
Field

**Figure 6.11 Fleming's left-hand rule shows how field, current and motion are connected.**

Figure 6.11 shows how the three directions (field, current and motion) can be identified using Fleming's **left-hand rule**. The motion is always at right angles to the direction of both the magnetic field and the current. Any movement is greater if:

- the magnetic field is stronger
- the current is higher
- there is a greater length of wire in the field.

Some ammeters are designed to show how much current is flowing in a wire by measuring the amount of movement caused by the motor effect.

## Deflection of charged particles

REVISED

Although the motor effect is usually observed because a wire moves, the force acts on the moving charges **within** the wire. If charged particles are moving at right angles to a magnetic field, then there is a current and a force will act to cause deflection. If the motion is parallel to the magnetic field, nothing happens, just as in the case of a wire that runs parallel to the field lines.

The size of the force is increased if:

- the magnetic field is stronger
- the particles are moving faster
- the particles have a greater charge.

> **Left-hand rule**: A way to work out the connection between magnetic field (**F**irst finger, **F**, **F**ield) current (se**C**ond finger, **C**, **C**urrent) and motion (thu**M**b, **M**, **M**otion)

### Exam tip

Working out the direction of a deflection is tricky. You need to remember that for the left-hand rule the direction is for **conventional** current (positive to negative, see page 32). This means the rule also works for positively charged particles like protons. But you need to reverse the direction if you are considering particles that have a negative charge, such as electrons.

## Now test yourself

TESTED

11  An electric drill has a motor to make it spin. What will happen to the drill if the current is increased?
12  List the three directions for the left-hand rule.
13  The Earth's magnetic field is very weak but if a high current flows in a wire it can still experience a force because of it. Which way must the wire be pointed for the motor effect to occur?
14  Newton's third law (see page 15 for a reminder) suggests that when there is a force on the wire there must be an equal and opposite force on something else. What is the other force exerted on and why does it not move?

Answers on page 137

# Electric motors

## The principle of the d.c. motor

Instead of one wire, a motor uses a coil of wire to increase the length of wire that is in the magnetic field. More loops in the coil give the electromagnet a stronger field, and this means a greater force. The coil in a motor is mounted so it spins within the permanent magnetic field.

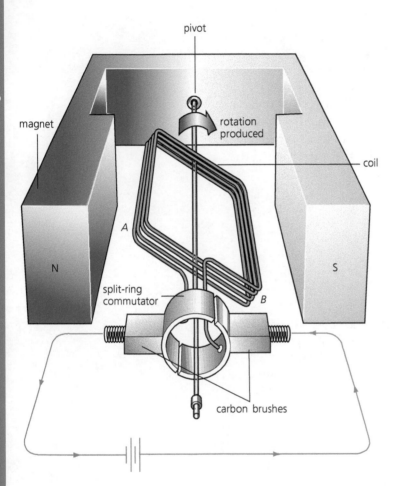

pivot

rotation produced

magnet

coil

A

N

S

split-ring commutator

B

carbon brushes

**Figure 6.12 A simple motor showing the coils of wire**

The motor effect can start the movement of a coil of wire, but as soon as the wires are parallel to the magnetic field there will be no turning force. Even if friction is low enough for the coil to complete more than a quarter turn, the force would now be in the opposite direction. To solve this problem, a **split-ring commutator** is built into the electric motor. The easiest way to understand this process is to break it down into steps, to consider both the current and the force. When the conductor of the ring touches the brushes, the circuit is complete and there is a temporary magnetic field around the coil. When the brushes are against the gap in the ring (the 'split'), the circuit is incomplete.

> **Split-ring commutator**: A component in a d.c. motor that spins with the coil, so the force on it due to the motor effect is always in the same direction

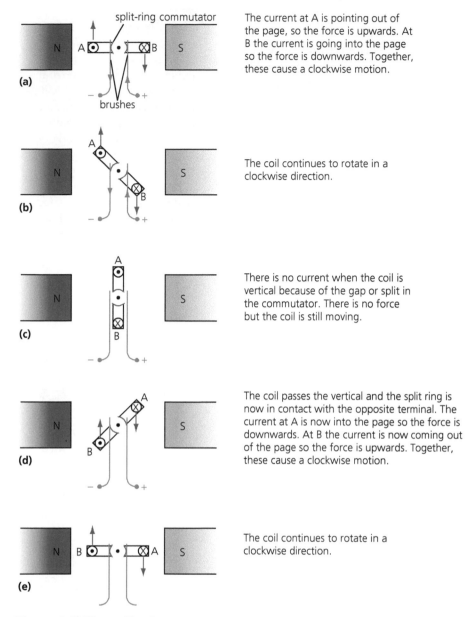

The current at A is pointing out of the page, so the force is upwards. At B the current is going into the page so the force is downwards. Together, these cause a clockwise motion.

(a)

The coil continues to rotate in a clockwise direction.

(b)

There is no current when the coil is vertical because of the gap or split in the commutator. There is no force but the coil is still moving.

(c)

The coil passes the vertical and the split ring is now in contact with the opposite terminal. The current at A is now into the page so the force is downwards. At B the current is now coming out of the page so the force is upwards. Together, these cause a clockwise motion.

(d)

The coil continues to rotate in a clockwise direction.

(e)

**Figure 6.13 The split-ring commutator**

If the current is in the opposite direction, the motor will spin anti-clockwise instead.

## The moving-coil loudspeaker

Sound waves are caused by a vibrating object, one that moves back and forth, causing the air to do the same (see page 55 for more information).

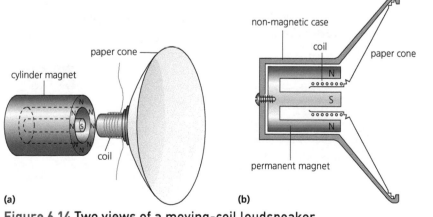

(a)

(b)

**Figure 6.14 Two views of a moving-coil loudspeaker**

> **Exam tip**
>
> You can work out two of the three directions (current, magnetic field and motion) using the left-hand rule, if you are given one of the directions in a diagram. Remember that you may need to check the directions on two parts of a coil.

Although the arrangement of the coil and magnet looks different from a d.c. motor, the loudspeaker also moves because of the motor effect. The changing size and direction of the alternating current means the coil experiences a changing size and direction of force. These changes cause sound waves of specific amplitude (loudness) and frequency (pitch). The direction of the force can be worked out with the left–hand rule.

## Now test yourself

TESTED

15  A student builds a motor, arranging the coils so that they are vertical in between the magnetic poles. What will happen when the battery is connected?
16  Two motors have the same strength magnets and are connected to identical power supplies, but one turns faster than the other. Explain, using the motor effect, why this might happen.
17  The sound from a loudspeaker decreases in pitch (frequency) and gets louder. Describe how the current changes to cause this effect.

Answers on pages 136–7

Answers on pages 136–7

> **Revision activity**
>
> On pieces of paper, write the names of some electrical devices that have a moving part. Some should be mains-powered and others should use batteries. Revise your knowledge by choosing one at random and describing the effect of changing the current on how the device works. Use the motor effect as part of your description.

# Electromagnetic induction

## Induced voltage

REVISED

When a conducting wire moves through a magnetic field, a voltage difference is produced; this is called an **induced voltage**. If the wire is part of a circuit, this will cause an induced current to flow. A voltage will also be induced if a stationary wire is in a changing magnetic field. This is called the generator effect.

> **Induced voltage**: A voltage that is produced by a wire which is subjected to a changing magnetic field, or that is moved within a magnetic field. The voltage causes a current.

## Investigating induced voltages

REVISED

Connecting the two ends of the wire to a sensitive voltmeter makes it easy to investigate the factors that affect the size and direction of the induced voltage. A voltage is only induced when the wire is moving at right angles to the magnetic field, so that the electrons experience a force along the wire. As with the motor effect, the three directions must all be at right angles to each other.

The factors that affect the size and direction of a voltage include:
● the speed of motion of the wire – increasing this creates a larger voltage
● magnet strength – using a stronger magnet creates a larger voltage, even if the speed of motion is the same
● direction of motion – reversing this reverses the direction of the voltage
● direction of the magnetic field – reversing this reverses the direction of the voltage.

> **Exam tip**
>
> There are formulae that can be used to calculate the induced voltage, but all you need to understand is the factors that affect the direction and relative values.

## Coils and magnets

REVISED

When a coil of wire (a solenoid) is connected to a voltmeter, a voltage and current can be induced if a magnet is moved into or out of the coil. The direction of the induced voltage depends on which pole of the magnet goes into the coil first. The size of the voltage depends on:
● the strength of the magnet
● the speed of motion
● the number of coils in the solenoid.

## Now test yourself

18  During an investigation of electromagnetic induction, 0.24V is induced in a wire when it is moved upwards. What will the value be if the magnets are reversed and the motion is at the same speed, but downwards?

19  A student drops magnets, north-pole first, through a long solenoid which is clamped vertically. What effect will each of the following changes have on the induced voltage:
    (a)  a stronger magnet
    (b)  dropping the magnet south-pole first
    (c)  fewer coils in the solenoid?

20  A piece of iron the same size and weight as the magnet in Question 19 is dropped through the solenoid. What induced voltage will this produce?

Answers on page 137

# Generators

A d.c. motor has design features (current and a permanent magnetic field) to produce continuous movement in the same direction, based on the motor effect. An a.c. generator is designed to induce an alternating current based on the generator effect (motion in a permanent magnetic field). Our mains supply is an alternating current, and in the UK this has a frequency of 50 Hz.

## The a.c. generator (alternator)

In an a.c. generator, turning the coil moves the wire through a magnetic field. Only movement of the wire at right angles to the magnetic field lines produces a voltage. The maximum, or peak, voltage is produced when the coil is horizontal, and there is no induced voltage when the coil is upright. This means that the voltage produced increases and decreases, reversing direction halfway through each rotation as the wires on each side of the coil switch between upward and downward motion. Look at Figure 6.15, (i) through (iv), for examples showing the voltage induced at different coil positions.

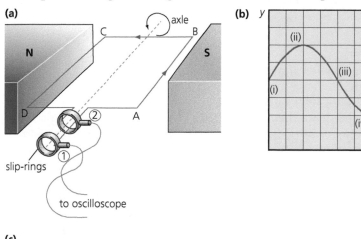

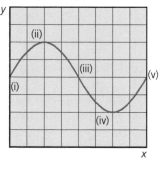

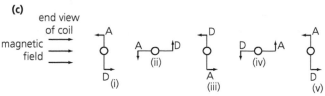

**Figure 6.15** For an a.c. generator (a), different voltages are produced (b) as the coil moves through different positions (c).

# The size of the induced voltage

The value given to describe an a.c. voltage is usually slightly less than the peak voltage, to make up for the times when the voltage is closer to zero (see page 22 for more information). The induced voltage is greater when:

- the coil rotates faster
- the magnet is stronger
- there are more turns of wire in the coil
- the coil has a soft iron core at the centre.

**Exam tip**

If the coil is rotated faster, there are two effects. As well as an increase in the maximum voltage, the time between each peak is also reduced. This is measured as an increase in the frequency.

# Producing power on a large scale

Almost all mains electricity is produced by large-scale devices based on the principle of an a.c. generator (the only exception being solar cells). When producing mains electricity a different method is used to provide the movement of the turbines (see page 70 for more information).

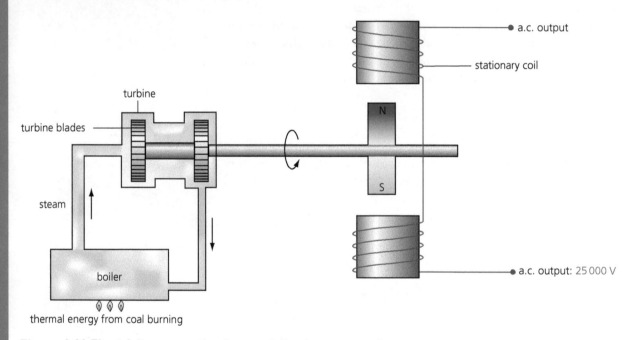

**Figure 6.16 Electricity generation in a coal-fired power station**

The magnet is on a rotating axle so that the output cables can be fixed in place. The burning coal boils the water and the steam then turns a turbine. This is what causes the axle and the magnets to turn, inducing a high voltage in the output wires.

**Exam tip**

Although the arrangement of components may seem different from the small-scale a.c. generator, the important thing is that the coils and magnets are moving relative to each other. Current is produced whether it is the coils or the magnets that are moving.

## Now test yourself

TESTED

21 Why does an a.c. generator not need a split-ring commutator?
22 A student explains that doubling the rotation speed will double the peak voltage of an a.c. generator. What other change will occur?
23 When is the largest voltage induced in the coil? You may wish to use a sketch in your answer.

Answers on page 137

# Transformers

## Changing fields and changing currents

Switching a current on or off in a solenoid causes a changing magnetic field. If another coil of wire is close by, this changing magnetic field induces a voltage in the wire (just as if a magnet was moving nearby). The induced voltage occurs only while the magnetic field is changing, which happens when the current is changing. A constant current causes a magnetic field, but does not cause an induced voltage in the second solenoid.

## Transformers

If two coils share an iron core, the changing magnetic field around the first coil has a greater effect on the second. Energy is transferred between the two coils even though they are not electrically connected. This arrangement is called a **transformer**.

In a transformer, there is a clear sequence of events involving electromagnetism and induction:

1 The primary coil is connected to an alternating current supply.
2 The changing current causes a changing magnetic field around the soft iron core.
3 An alternating voltage is induced in the secondary coil because it is in a changing magnetic field.

If the number of turns of wire on the secondary coil is greater than on the primary coil, the output voltage will be increased. If there are fewer turns on the secondary compared to the primary, the output voltage is decreased. A **step-up transformer** is used to increase voltage for its transmission over large distances of many kilometres. A **step-down transformer** might be used to reduce mains voltage for household devices.

> **Step-up transformer**: A transformer with more turns on the secondary coil, so the secondary voltage is increased compared to the primary voltage
>
> **Step-down transformer**: A transformer with fewer turns on the secondary coil, so the secondary voltage is decreased compared to the primary voltage

The ratio of the turns on the primary and secondary coils is the same as the ratio of the voltages. For example, if the number of turns doubles, so does the voltage.

$$\frac{\text{input (primary) voltage}}{\text{output (secondary) voltage}} = \frac{\text{input (primary) turns}}{\text{output (secondary) turns}}$$

$$\frac{V_P}{V_S} = \frac{N_P}{N_S}$$

The subscripts $P$ and $S$ stand for primary and secondary. The primary coil is the input and the secondary coil is the output.

> **Transformer**: Two coils of wire on a soft iron core. Energy is transferred between the primary and secondary coils because of electromagnetism and electromagnetic induction.

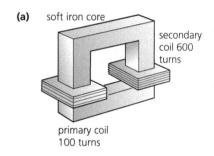

(a) soft iron core

secondary coil 600 turns

primary coil 100 turns

(b)

**Figure 6.17** (a) A step-up transformer and (b) the circuit symbol for a transformer

> voltage, $V$, measured in volts (V)
>
> number of turns on the coil, $N$, which is a number with no units

## Example

An a.c. supply, 2V, is connected to the input of a transformer with 100 turns on the primary coil and 600 turns on the secondary coil. What is the output or secondary voltage?

Answer

$$\frac{V_P}{V_S} = \frac{N_P}{N_S}$$

$$V_S = \frac{N_S}{N_P} \times V_P$$

$$V_S = \frac{600}{100} \times 2$$

$$V_S = 6 \times 2$$

$$V_S = 12\,V$$

## Exam tip

Sometimes the terms primary or secondary are used to identify coils and values. Sometimes they are described as input or output instead. If using symbols with subscripts, you may prefer to use 1 and 2 rather than letters.

## Exam tip

You might find this calculation easier to do using ratios, through looking for an easy multiple shared between the primary and secondary values. Make sure you check your answer, and remember that voltage and turns both increase for a step-up transformer, and both decrease for a step-down transformer.

## Typical mistake

Make sure your answers don't suggest or imply that the primary and secondary coil are connected electrically – they are two separate circuits. It is the electromagnetic effect that links them together.

# Power in transformers

REVISED

The power transferred by electricity is calculated using the current and the voltage (see page 25 for more information). For a transformer that is 100% efficient the power transferred by the primary coil must be equal to the power transferred by the secondary coil.

$$V_P \times I_P = V_S \times I_S$$

voltage, $V$, measured in volts (V)

current, $I$, measured in amperes or amps (A)

## Example

A step-up transformer increases the voltage from 2V to 12V. The current in the primary coil is 1.8A. What is the output or secondary current?

Answer

$$V_P \times I_P = V_S \times I_S$$

$$I_S = \frac{V_P \times I_P}{V_S}$$

$$I_S = \frac{2 \times 1.8}{12}$$

$$I_S = \frac{3.6}{12}$$

$$I_S = 0.3\,A$$

For a step-up transformer, the voltage increases and the current decreases. For a step-down transformer, the voltage decreases and the current increases.

# The National Grid

Step-up transformers are used to increase the voltage (and decrease the current) before transmission over long distances. This is because a high current would cause heating of the wires, and energy would be dissipated to the surroundings instead of being transferred usefully. Step-down transformers are then used to reduce the voltage so that it is safe (see page 38 for more information).

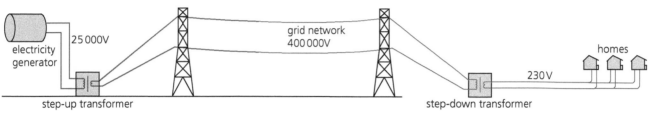

**Figure 6.18 The National Grid uses step-up and step-down transformers.**

## Revision activity

The ideas in this topic link many concepts, from forces and motion to voltage and current. For each key idea, review how it relates to electromagnetism. A mind map may be a good way to show the links without too much writing.

## Now test yourself

24  A step-down transformer in a local sub-station reduces the voltage from 11 500 V to 230 V. There are 5000 turns on the secondary coil. How many coils would be on the primary coil?

25  In a school lab, a transformer has 230 V supplying 0.2 A through the primary coil. The secondary current is 2 A. What is the secondary voltage, assuming there is 100 % efficiency?

26  Why can't transformers be used with direct current?

Answers on page 137

## Summary

- Magnets have a north-seeking and a south-seeking pole. Opposite poles attract and like poles repel.
- Magnetic fields are shown with field lines that point from north poles to south poles. The field is strongest (and the force greatest) where the lines are closest together.
- When iron, nickel and cobalt (magnetically soft materials) are placed in a magnetic field magnetism is induced. Steel is magnetically hard and once magnetised it becomes a permanent magnet.
- An electric current in a wire produces a temporary magnetic field around it. The right-hand grip rule predicts the direction of the field.
- A coil of wire, called a solenoid, has a magnetic field similar to that of a bar magnet. The field is

stronger when the current is higher, when there are more turns of wire in the coil or if there is a soft iron core.
- A current-carrying wire experiences a force in a permanent magnetic field and may move. This is called the motor effect. The left-hand rule links the directions of current, magnetic field and motion.
- Motors are designed to rotate when current flows in the coil. The speed will be higher if the current is higher, when there are more turns of wire in the coil or if the magnet is stronger.
- A wire moving at right angles in a magnetic field, or in a changing magnetic field, will have an induced voltage. This is called the generator effect and is used to generate mains electricity. The induced voltage will be higher if the motion is faster, when there are more

6 Magnetism and electromagnetism

turns of wire in the coil or if there is a soft iron core.
- Transformers have a primary coil and a secondary coil which share a soft iron core. They increase or decrease the voltage depending on the relative number of turns of wire on each coil.

$$\frac{V_P}{V_S} = \frac{N_P}{N_S}$$

- The input and output power will be the same on a transformer which is 100% efficient.

$$V_P \times I_P = V_S \times I_S$$

- The National Grid uses step-up transformers to increase voltage (for reduced energy dissipation over long wires) and step-down transformers to decrease voltage (for safe consumption).

## Exam practice

1 Students are investigating data on the output from a coal-fired power station.
  (a) Which of these is not used as a source of heat in a thermal power station? [1]
    A  oil            C  geothermal
    B  uranium        D  solar cell
  (b) State what turns the turbine in this kind of power station. [1]
  (c) The output voltage is shown in the figure below.

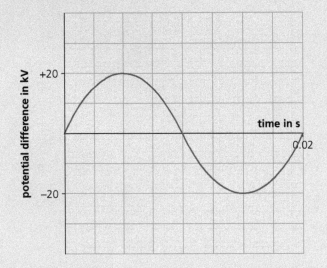

    (i) Identify the time at which the coil is at right angles to the magnetic field. [1]
    (ii) Describe how the oscilloscope trace would change if the turbines were spinning twice as fast. [2]
  (d) The voltage produced by the turbine is used as the input for a transformer connected to the National Grid. Describe the type of transformer that would be used and give reasons for your answer. [3]

2 A teacher is demonstrating the effects of magnets and electromagnets.
  (a) The pattern in the figure below is demonstrated with two magnets and iron filings.

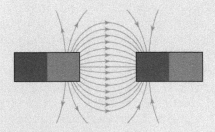

    (i) Describe the poles of the magnets based on the pattern in the figure. [1]
    (ii) Identify what kind of field exists between the magnets. [1]

(b) State what force you would expect between the magnets in this position. [1]

(c) A solenoid is set up with a current flowing. Describe how the magnetic field lines could be sketched using plotting compasses. [3]

(d) State how the field lines would be different if

    (i) a soft iron core was placed at the centre of the solenoid [1]

    (ii) the direction of the current was reversed. [2]

3 A transformer is used to decrease the voltage from 230V for a bathroom shaver socket.

(a) There are 575 turns on the primary coil and 30 turns on the secondary coil. Show that the secondary voltage is about 12V. [3]

When a device that is plugged into the socket is working normally, the current flowing through it is 0.45A.

(b) (i) Calculate the current in the primary coil. [3]

    (ii) What assumption have you made for this calculation? [1]

(c) Explain why transformers only work with an alternating current supply. [2]

4 Proton beam therapy uses positively charged particles travelling at high speed to kill cancer cells.

(a) A strong magnetic field is needed to deflect the particles so that they can be aimed. Give two reasons why an electromagnet is better than a permanent magnet for this task. [2]

(b) Explain why the electromagnet is cooled with liquid nitrogen. [2]

(c) (i) Using the above figure, state whether the protons are deflected in an upwards or downwards direction. The arrow shows the original direction of the protons. [1]

    (ii) Explain your answer. [1]

(d) Electrons have a negative charge and a smaller mass. Describe how the direction and size of deflection would change. [2]

5 Students are investigating the design of an electric motor.

(a) Explain what will happen to the coil of wire as soon as the current is switched on. [3]

(b) (i) Identify component X. [1]

    (ii) Explain how X allows the coil to turn continuously. [2]

(c) Suggest two changes to the circuit that would increase the speed of the motor. [2]

**Answers and quick quizzes online**

ONLINE

# 7 Radioactivity and particles

## Atomic structure

### Neutrons, protons and electrons

Each atom is made up of **protons** and **neutrons** that exist in a very small nucleus, with **electrons** orbiting around it. Almost all the mass of an atom is in the nucleus, which is positively charged because of the protons there. Electrons have almost no mass and are negatively charged. Neutrons are uncharged or neutral.

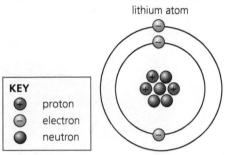

**KEY**
- ⊕ proton
- ⊖ electron
- ● neutron

**Figure 7.1 Protons, neutrons and electrons in a lithium atom (Not to scale)**

Each atom is mostly empty space, with the nucleus around 10 000 times smaller than the distance out to the orbiting electrons. Atoms have neutral charge overall because they have equal numbers of protons and electrons.

> **Proton**: A subatomic particle with a +1 (positive) charge and a relative mass of 1, found in the nucleus of an atom
>
> **Neutron**: A subatomic particle with no charge and a relative mass of 1, found in the nucleus of an atom
>
> **Electron**: A subatomic particle with −1 (negative) charge and a negligible mass, found orbiting the nucleus

### Ions

If an atom gains or loses electrons, the positive and negative charges are no longer equal. This means the **particle** is now an ion. If it gained electrons it will be a **negative ion**. If it lost electrons it will be a **positive ion**.

> **Particle**: A single unit of a material which could be an atom, ion or molecule. It is also used for parts of the atom (protons, neutrons and electrons), and the products of nuclear decay emitted from the nucleus (alpha, beta and neutrons).
>
> **Negative ion**: An atom that has gained orbiting electrons and so has an overall negative charge. These are shown by including a negative value after the symbol (for example Cl⁻).
>
> **Positive ion**: An atom that has lost orbiting electrons, and so has an overall positive charge. They are shown with a positive value after the symbol (for example $Ca^{2+}$).

> **Exam tip**
>
> In chemistry, a positively charged ion is sometimes called a cation and a negatively charged ion is an anion, because of the electrodes they are attracted to.

## Atomic and mass number

A chemical element is determined by the number of protons in the nucleus. This is called the **atomic number** or **proton number**. All atoms of an element have the same number of protons. For example, all boron atoms have five protons.

The number of neutrons in an element is not fixed, although some combinations are more common than others. The **mass number** or **nucleon number** is the total number of protons and neutrons in the nucleus (a nucleon is a proton or a neutron). For any single atom this must be a whole number, but the values given in the Periodic Table are averages. By knowing the mass number and the atomic number for an atom, the number of each type of subatomic particle can also be worked out.

$$^{11}_{5}B$$

- Atomic number = 5: this means there are 5 protons.
- Mass number = 11: this means there are $11 - 5 = 6$ neutrons.
- For an atom the number of electrons and protons are equal: this means there are 5 electrons.

> **Exam tip**
>
> The elements in the Periodic Table are listed in order of their proton number.

> **Atomic number** or **proton number**: The number of protons in the nucleus of an atom
>
> **Mass number** or **nucleon number**: The total number of protons and neutrons in the nucleus of the atom, and the mass in relative units

## Isotopes

Although every atom of an element has a fixed number of protons, the number of neutrons is more variable. These different forms are chemically identical but have different masses and are called **isotopes**. For example, chlorine exists in two common stable isotopes (as well as in other unstable ones which are much rarer) called chlorine-35 and chlorine-37.

$$^{35}_{17}Cl \qquad ^{37}_{17}Cl$$

> **Revision activity**
>
> Many of these ideas crossover to chemistry. Make summary notes about the ideas here, perhaps with a column for differences in language between the subjects. The numbers of protons and electrons are important in chemistry because of chemical bonding, while the neutron number and the isotopes may be more relevant in physics.

> **Typical mistake**
>
> Take care with similar looking words when working under pressure in an exam. Mixing up neutron, nucleon and nucleus will lose you marks.

> **Isotopes**: Atoms of the same element that have different numbers of neutrons and so different masses

> **Typical mistake**
>
> It is not just the rare or unstable atoms that have isotopes. For example, the three different types of carbon atoms – all with six protons but with varying numbers of neutrons – are isotopes.

### Now test yourself

TESTED

1. (a) How many protons and neutrons are there in the lithium atom shown in Figure 7.1?
   (b) What is its atomic number?
   (c) What is its mass number?
2. Explain the difference between an ion and an isotope.
3. A rare isotope of carbon is written as $^{14}_{6}C$. How is this different to the more common isotope, which can be written as carbon-12?

Answers on page 137

# Radioactivity

## Nuclear decay

Most atoms are stable, with a nucleus that does not change. Some nuclei are less stable and emit particles to become more stable. This is called nuclear decay and the numbers of the subatomic particles in the nucleus change during this process. Sometimes electromagnetic radiation is emitted as well. All these different emissions are called **radioactivity**.

Nuclear equations show what is present at the start and end of nuclear decay, and are written so that the total masses and charges are the same before and after the emission. There are four types of radioactive emission you need to understand and remember. These are:
- alpha particles
- beta particles
- gamma rays
- neutron radiation.

### Alpha particles

**Alpha particles** are made up of two protons and two neutrons, the same as a helium nucleus. They have a +2 charge and a mass of four. They are usually only emitted from very heavy elements, and the nucleus left behind has two fewer protons and two fewer neutrons, so it becomes a different element. This is called **alpha decay**.

$$^{238}_{92}U \rightarrow ^{234}_{90}Th + ^{4}_{2}He$$

uranium nucleus → thorium nucleus + alpha particle (helium nucleus)

### Beta particles

**Beta particles** are fast-moving electrons. They are emitted from a nucleus when a neutron changes into a proton (which stays) and an electron (which does not). The nucleus left behind therefore has one fewer neutron and one extra proton, but the mass change is too small to measure. The proton number of the nucleus increases by one, so it becomes a different element. This is called **beta decay**.

$$^{14}_{6}C \rightarrow ^{14}_{7}N + ^{14}_{7}N$$

carbon nucleus → nitrogen nucleus + beta particle (electron)

### Gamma rays

**Gamma rays** are sometimes emitted after a nucleus has emitted an alpha or beta particle. These rays are a kind of electromagnetic radiation, not a particle. They do not change the subatomic particles in the nucleus which is why this process is not always called decay (see page 49 for more information about gamma rays).

### Neutron radiation

**Neutrons** are sometimes emitted from highly unstable nuclei. This means the mass number is reduced by one but it does not become a different element. Neutron radiation is highly dangerous but is also very rare.

$$^{5}_{2}He \rightarrow ^{4}_{2}He + ^{1}_{0}n$$

helium 5 nucleus → helium 4 nucleus + neutron

**Radioactivity**: The emission of particles or electromagnetic radiation from a nucleus following nuclear decay

**Exam tip**

As there are different kinds of emissions, it is best to be specific. The term 'radiation' by itself may be unhelpful as it can apply to electromagnetic radiation, which is not always emitted from a nucleus.

**Alpha particle**: Two protons and two neutrons, emitted from an unstable nucleus; written as $^{4}_{2}\alpha$ or $^{4}_{2}He$

**Alpha decay**: The emission of an alpha particle from an unstable nucleus

**Beta particle**: Fast-moving electron, released by the breakdown of a neutron in an unstable nucleus to a proton and electron; written as $^{0}_{-1}\beta$ or $^{0}_{-1}e$

**Beta decay**: The emission of a beta particle from a nucleus

**Gamma rays**: Electromagnetic radiation with a very short wavelength and a high frequency; written as $\gamma$

**Neutrons**: A rare form of radioactivity which is highly dangerous; written as $^{1}_{0}n$

# Ionisation

All four forms of radioactivity can be described as ionising radiation. This means that when they collide with or are absorbed by materials they can cause **ionisation**. The ions produced when the radiation is absorbed can cause damage in living tissue including changes to DNA. These changes to living tissue are called mutations and can lead to cancer.

Some kinds of nuclear radiation are more ionising than others:
- Alpha radiation is very strongly ionising.
- Beta radiation has medium ionising power.
- Gamma radiation is weakly ionising.
- Neutron radiation causes ionisation indirectly, by causing other atoms to become unstable.

> **Ionisation:** The production of ions by radioactivity due to the energy transferred

> **Exam tip**
>
> Make sure your answers do not suggest that it is the charge of the particles which causes ionisation. The different kinds of radioactivity are all ionising because they transfer energy during collision or absorption.

> **Revision activity**
>
> A Venn diagram is a useful way to summarise the different kinds of radioactivity. Using this technique means you have to consider the similarities and differences between each kind, and how these characteristics affect their behaviour.

## Now test yourself

4 Explain why an alpha particle is sometimes referred to as a 'helium nucleus' and sometimes as '2p+2n'.
5 Copy and complete the summary table for the different types of radioactivity.

| | Emitted from nucleus | Change in mass number | Change in atomic number |
|---|---|---|---|
| Alpha | 2p + 2n | | |
| Beta | | 0 | |
| Gamma | | | 0 |
| Neutron | | | |

6 Which kind(s) of radioactivity:
   (a) cause ionisation
   (b) have no charge
   (c) decay to leave a different element?

7 $^3_1$H decays by emitting a beta particle. Write the nuclear decay equation to work out the proton and mass numbers of the new nucleus.

Answers on page 137

# The nature of alpha, beta and gamma radiation

## Detecting particles

Initially, the different kinds of nuclear radiation were described by their effects. For example, they can be detected by their effect on photographic film (the radiation has the same effect as visible light), or by their ionising properties. A detector called a Geiger-Müller (GM) tube is used to measure the ionisation of atoms in a gas caused by radiation, but it cannot distinguish between the different types.

## Required practical

### Investigate the penetration powers of different types of radiation using radioactive sources

#### Method

Alpha, beta and gamma sources can be used by a trained teacher in school. Neutron sources are too dangerous for classroom use.

1 The teacher set up the Geiger-Müller tube, connected to a counter that recorded each detection. Students stood at a safe distance.
2 In turn, each radioactive source was lifted with long tongs and pointed towards the GM tube.
3 The number of detections in a minute was recorded for each source, through air and through different materials. Each measurement was repeated three times.

Although the pattern was clear, the exact number of detections varied. The average results were used to compare the penetration power of each kind of nuclear radiation.

#### Results

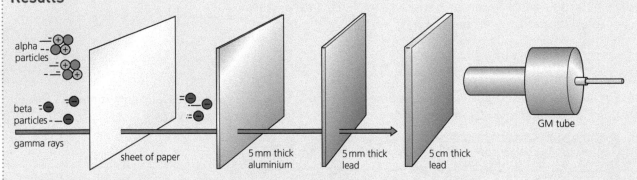

**Figure 7.2 Penetration of alpha, beta and gamma radiation**

- The alpha radiation was stopped by a sheet of paper or a few centimetres of air.
- The beta radiation went through the paper and did not decrease much over several metres of air. It was stopped by a thin sheet of aluminium.
- The gamma radiation went through the paper and the aluminium and showed no change in air. It was reduced by a thin layer of lead and stopped by a thick layer of lead.

## Background radiation

REVISED

We are all exposed to nuclear radiation from the environment. Some is natural and some is caused as a result of human activity. The amount of **background radiation** in the area depends on the place. For example, some radioactive rocks release radon over time, which can be inhaled if it builds up in cellars or underground. People who live at altitude or spend a lot of time flying in planes are exposed to more cosmic rays. In most places artificial sources, such as nuclear power stations or the remains of nuclear weapons testing, only make up a very small amount of the background radiation.

People who work with radioactive materials, for example in nuclear reactors or in some medical facilities, take extra precautions to reduce their exposure.

> **Background radiation:** The nuclear radiation from natural and artificial sources in our everyday environment

> **Exam tip**
>
> Remember that background radiation is an average, and local areas might have more or less. In Britain the biggest variable is the underlying geology as some rocks are more radioactive than others.

TESTED

## Now test yourself

8 An unknown sample is tested with a Geiger-Müller tube. The count rate went down to nearly zero when there was aluminium in between the source and detector, but was not affected by a piece of paper. What was the source emitting?

9 Why might 'airline pilot' and 'coal miner' both be careers that involve exposure to nuclear radiation?

10 A factory makes smoke alarms using tiny amounts of an alpha radiation source. Why is it more important for the staff to wear a face mask than to wear gloves while they work?

11 What material is used as shielding for radioactive samples in school?

Answers on page 137

# Radioactive decay

## Random process

REVISED

Nuclear radiation is emitted when an unstable nucleus decays. This process is predictable for a large sample. For single atoms, it is a random process and impossible to predict.

## Decay

REVISED

On average, a sample emits less radiation over time. This is because when an unstable atom decays it leaves behind a smaller sample of atoms that can decay. The number of atoms that decay each second is called the **activity** and is measured in **becquerels** (Bq). The higher the activity, the quicker the unstable atoms decay and the faster the sample will stop being radioactive.

> **Activity**: The number of atoms that decay each second. The more unstable the isotope, the higher the activity will be. The count rate of a Geiger-Müller tube is often used to estimate the activity, which reduces over time for a sample.
>
> **Becquerels (Bq)**: The unit of activity and a way to describe how much radiation a sample or source is emitting. Often units of milli-becquerels (mBq) are used.

## Measurement of half-life

REVISED

If measurements of activity are recorded for a sample, a pattern can be seen. For each different isotope there is a fixed amount of time during which the average activity is halved. This is called the **half-life**, $t_{\frac{1}{2}}$. For some isotopes, it takes millions of years. For others, it takes only seconds. The more stable an isotope, the longer the half-life will be.

As you have just read, after one half-life, however long that is for a radioactive isotope, the activity will be halved. After two half-lives it will have halved again, which means the activity will be a quarter of the **original**. The half-life for an isotope is not affected by temperature or chemical reactions.

> **Half-life**, $t_{\frac{1}{2}}$: The amount of time taken for the activity of a sample, and the number of remaining unstable atoms, to be halved. This should be measured in seconds, but for very long half-lives it may be given in other units for convenience.

### Typical mistake

Do not suggest that there are fewer atoms after nuclear decay. The total number of atoms has not changed, but the number of **unstable** atoms reduces over time, which is why the average activity decreases.

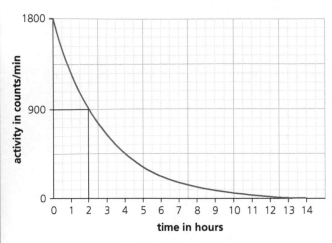

**Figure 7.3 The radioactive decay curve for a substance with a half-life of 2 hours**

If activity is plotted against time, lines can be drawn to show the half-life, as shown in Figure 7.3.

**Exam tip**

A half-life is specific to the isotope, not the element. Carbon-12 is so stable the half-life can be treated as infinite. Carbon-11 has a half-life of 20 minutes and carbon-14 has a half-life of 5700 years.

## Dating archaeological remains

REVISED

All living things take in carbon from the environment, either as food or as carbon dioxide. Some of this will be carbon-14. If the activity of an historical sample that contains carbon can be measured and compared to a similar modern sample, the reduction in activity can be used to tell us how old it is.

**Example**

A sample of preserved wood has an activity one-eighth of the original value (worked out by comparing it with a similar but modern sample). How old is the sample? (Carbon-14 has a half-life of 5700 years.)

Answer

- After one half-life or 5700 years the activity will be half the original value.
- After two half-lives or 11 400 years the activity will be a quarter of the original value.
- After three half-lives or 17 100 years the activity will be an eighth of the original value.

So the sample is around 17 100 years old.

**Revision activity**

Make a list of some of the key words in this section and check that you can recall the meaning of each one by testing yourself daily. Mix up the order to ensure it is the definition you are learning and not just remembering the order of terms. As you become more confident, pick two words at random and explain how they could be related to each other.

## Now test yourself

TESTED

12 Two radioactive samples are tested by a teacher and have the same activity of 2000 Bq. One has a half-life of 2 days, the other of 2 years. Explain what you would expect to find in each sample when the measurement is repeated 4 days later.
13 Will the half-life be large or small for a very unstable isotope?
14 How is the measured count rate for a sample different to the activity?

Answers on page 137

# Uses of radioactive materials

## Medicine

Because ionising radiation can kill living cells, it is used to sterilise equipment. Intense gamma sources such as cobalt-60 are aimed at sealed containers of surgical tools, dressings and syringes. This will kill any micro-organisms, such as bacteria, without the complications associated with heating or use of chemicals such as bleach.

There are two separate ways that radioactivity is used directly in medicine – for diagnosis and treatment. As with sterilising equipment, cobalt-60 and similar sources are used to deliberately kill living cells, in particular cancer cells within the body. The dose and position of the emitted radiation is very carefully planned, but some healthy cells are always affected and the side effects can be unpleasant. This treatment is called radiotherapy.

Many other conditions can be investigated with radioactivity, often in combination with X-ray imaging (see page 49 for more information), as a form of diagnosis. Because the emitted radiation can be detected outside the body, a chemically safe **tracer** is injected, inhaled or swallowed. Tracers with short-half-lives are used, so the person only receives a low dose, but while radiation is being emitted the radiographer can use X-rays to tell how well the target organ is working. For example, if radiation from an inhaled gas is only detected from one side of the chest it suggests there is a blockage in the lung.

> **Tracer**: A radioactive isotope, usually a gamma emitter with a short half-life, that is used to detect the position of leaks or blockages in systems that are hard to investigate. These can include underground pipes or human body systems.

## Industry

Tracers are used in other settings apart from medicine, such as in industry. For example, if a radioactive material is placed inside a container or pipe, a Geiger-Müller tube can be used to detect increases in emitted radiation. This will show where leaks are happening, even if they are too small to detect with other methods. Again, it is important to use sources with a short half-life.

## Irradiation and contamination

If a person is exposed to radiation from a radioactive source, they have been irradiated. If the activity and penetration of this radiation is high, damage may be caused. However, as soon as the source is removed, the irradiation stops. In some medical treatments, carefully aimed **irradiation**, used in controlled amounts, is necessary despite the possible side effects.

Accidental irradiation is a problem but rarely happens. If some of the radioactive source gets into a place where it is not wanted **contamination** has happened. This is a kind of pollution. The rare cases of reactor incidents have led to some contamination, both of the local area and then more widely as air and water carry the isotopes around the world.

Contamination is particularly dangerous when some of the radioactive source gets into the body. This is hard to remove and so continues to irradiate living cells, leading to health problems or even death.

> **Irradiation**: Exposure to nuclear radiation. The irradiation stops when the source is removed or blocked.
>
> **Contamination**: When the source of radiation is found somewhere it is not wanted, usually because of an accident. This means that people may be exposed to nuclear radiation, sometimes without knowing.

> **Revision activity**
>
> Review the practical method for measuring radioactivity and identify the ways in which both irradiation and contamination are limited. Make sure you can explain the precautions those working with radioactive sources need to take to avoid accidental irradiation and contamination.

TESTED

## Now test yourself

15 Which part of the body might be under investigation in a patient who is asked to swallow a radioactive tracer?
16 An engineer wants to check the thickness of aluminium foil in a factory, and so places a radioactive isotope underneath the conveyer belt and a detector above. What kind of radioactive source would work best and why?
17 It is thought an engineer might have accidentally inhaled some dust containing americium-241, which is an alpha emitter. Why would it be hard to test to see if they are contaminated?

Answers on page 137

# The hazards of radiation

The damage caused by radiation depends on the type of radiation, the ionising power and how much of the radiation reaches the living cells.

## Damage from different types of radiation

REVISED

Alpha sources are not dangerous if they are outside the body because they are stopped by air or the skin. If they get inside the body, for example by being inhaled or swallowed, they can be highly dangerous. This is partly because the radiation is strong enough to cause direct damage, colliding with cells and destroying body tissue. Neutron radiation from nuclear reactors does this too, but is penetrating enough to be damaging even from outside the body.

Alpha is also the most highly ionising form of radiation. This is a problem because it means alpha radiation is absorbed within short distance and ions form inside the body damaging cells and DNA. This is called **mutation** and can lead to cancer. Beta and gamma radiation are not so strongly ionising, so they are absorbed more gradually, which means the damage is spread out.

Gamma sources are dangerous, even though they are weakly ionising, because they always penetrate the body unless lead or concrete shielding are in the way.

The more radiation a person is exposed to, especially in a short time, the more likely it is they will have harmful effects. Although small doses are to be expected from background radiation, flights or medical scans, the risk of damage is small. Other lifestyle factors such as smoking, diet, traffic fumes and exercise contribute much more to health problems than nuclear radiation.

> **Mutation:** Damage to the DNA in a living cell. The cell may either die or start reproducing uncontrollably, which is known as cancer.

> **Exam tip**
>
> If answering a question about the effects of radiation on the body, make sure you give clear links between the properties of the radiation (ionising and penetration power) and the damage caused.

## Disposal of radioactive waste

REVISED

The waste from nuclear power stations contains a mixture of isotopes with a high activity (which makes them dangerous in the short term) and those with a long half-life (which makes them dangerous for many decades).

The waste from these stations is only produced in small quantities but the best storage methods and locations are highly debated. Usually waste is sealed in containers that use concrete and steel to prevent leaks, and then placed deep underground or at the bottom of the sea. Most scientists believe this is the safest method available but not everybody agrees.

# Nuclear fission

## Fission

REVISED ☐

During radioactive decay an unstable nucleus emits an alpha or beta particle, leaving behind a more stable nucleus that is slightly different. For some large nuclei the process leaves two roughly equal pieces, instead of emitting a much smaller particle, called daughter nuclei. When this happens, it is called **nuclear fission**. For uranium this usually happens when a neutron hits the nucleus. The exact size of the nuclei produced varies and there may be two or three neutrons emitted. An example of this reaction is:

$$_{0}^{1}n + _{92}^{235}U \rightarrow _{92}^{236}U \rightarrow _{52}^{137}Te + _{40}^{97}Zr + 2\,_{0}^{1}n$$

> **Nuclear fission**: A kind of nuclear decay when the unstable nucleus divides roughly in half instead of emitting a much smaller particle. Only a few elements undergo fission, such as uranium and plutonium, and it releases a lot of energy.

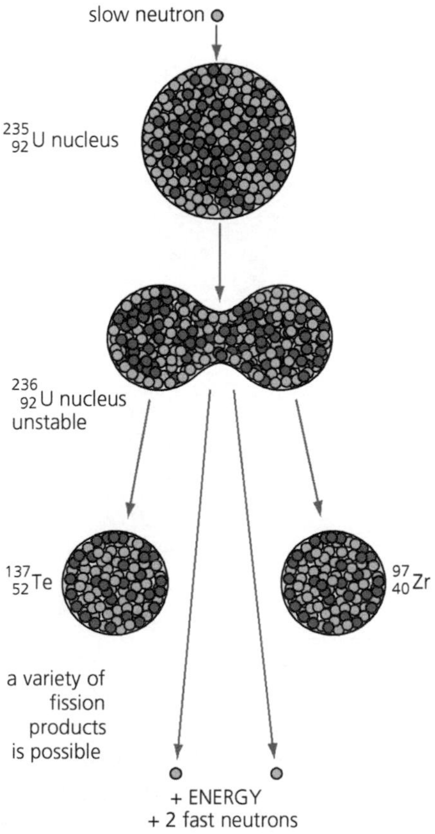

slow neutron ○

$_{92}^{235}U$ nucleus

$_{92}^{236}U$ nucleus unstable

$_{52}^{137}Te$          $_{40}^{97}Zr$

a variety of fission products is possible

○          ○
+ ENERGY
+ 2 fast neutrons

**Figure 7.4 The fission of a uranium-235 nucleus**

# Nuclear energy

REVISED

The daughter nuclei and neutrons created from fission have a lot of kinetic energy. A very careful measurement of the mass before and after the reaction shows that mass has been converted to energy, and the amount can be calculated by using the most famous equation in physics: $E = mc^2$ (although you don't need to know this for the exam). This is what is meant by the nuclear store.

In the right conditions, these nuclei and neutrons will heat up other nearby atoms. This means the energy can be used to heat water for steam, which then turns a turbine. This is similar to any coal-fired power station, but the fission of uranium produces far more energy than the combustion of coal or any other fossil fuel (see page 70 for more information). Energy is also released during nuclear decay, but as well as being a smaller amount than from fission, it is much less predictable and harder to control.

> **Typical mistake**
>
> When writing similar words, make sure your spelling is clear. For example, you should double check that you have spelt 'fission' and 'fusion' correctly and that you are using the right term in the right place. If the spelling is somewhere in between you may lose a mark for something you knew.

# Chain reaction

REVISED

Nuclear fission of uranium is such an effective source of heat for a power station because of the neutrons involved. Two or three are emitted when the first nucleus splits. If at least one of these is absorbed by another uranium-235 (U-235) nucleus, then that nucleus will split too, emitting more neutrons. This is called a **chain reaction**, and more and more uranium atoms start to undergo fission, releasing more and more energy and neutrons.

> **Chain reaction**: When neutrons from the fission of one uranium nucleus are allowed to cause fission in more uranium nuclei, which release more neutrons to continue the process

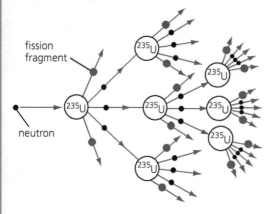

**Figure 7.5 A chain reaction in uranium-235**

If this process is controlled, by absorbing some of the neutrons in other materials, a chain reaction produces a predictable and controllable amount of heat. If all of the neutrons are allowed to continue the reaction without being controlled, the result is an explosion. This science is the basis of nuclear weapons.

# Nuclear power stations

REVISED

Rather than burning coal, a nuclear power station produces heat in a **nuclear reactor**. Instead of boiling water directly, most use a hot gas such as carbon dioxide which heats more predictably than water. Around the reactor layers of concrete shielding absorb the dangerous neutron radiation which would otherwise hurt the workers. Concrete is a particularly good absorber of neutrons because it contains water.

> **Nuclear reactor**: The heat source of a nuclear power station and where fission happens in a controlled way

There are three materials inside the reactor which are used to make sure the fission process happens at the right speed:

- The **fuel rods** contain uranium-235 (U-235), which is the starting point for fission, along with uranium-238 (U-238) which is the most common isotope of the element. Over time the uranium is used up, one atom at a time, as it breaks down into daughter nuclei which are themselves unstable. Eventually the fuel rods need to be replaced and the old ones are processed as radioactive nuclear waste (see page 71).
- The fuel rods are thin, so that the fast-moving neutrons can escape and travel through the **moderator** material, such as graphite. This slows down the neutrons so that they can be absorbed by uranium-235 nuclei in another fuel rod. As the neutrons slow down the gas is heated up.
- The **control rods** are very important, because they stop the chain reaction. They are made of boron, which absorbs the neutrons so they cannot cause more fission. They can be moved in or out of the gaps between the fuel rods to adjust the speed of the chain reaction, or be dropped in to stop it completely.

**Fuel rods:** Enriched uranium rods to supply nuclei for fission

**Moderator:** Graphite that slows down neutrons so they can be effectively absorbed to continue the chain reaction

**Control rods:** Rods made of boron that absorb neutrons so they can't carry on the chain reaction

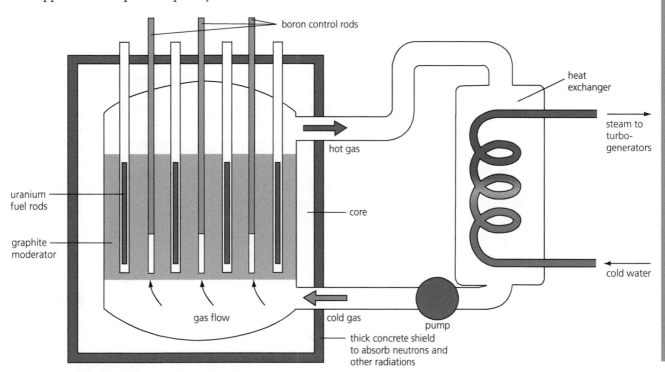

**Figure 7.6 A gas-cooled nuclear reactor.**

## Now test yourself

21 How are the nuclei of uranium-235 and uranium-238 different? Remember that the atomic number of uranium is 92.
22 The energy released by the fission process starts in the nuclear store of the uranium atoms. Describe the transfers that occur after that.
23 Explain the use of each of the following materials in a nuclear reactor:
   (a) boron   (b) uranium   (c) concrete   (d) graphite.

Answers on page 138

# Nuclear fusion

Energy is released when nuclei change to make a more stable arrangement. With nuclear decay and fission large nuclei make smaller ones. However, it is also possible for smaller atoms to join together to make more stable larger ones. This process is called **nuclear fusion**.

$$^3_1H + {}^2_1H \rightarrow {}^4_2He + {}^1_0n + \gamma \text{ ray}$$

When deuterium and tritium – isotopes of hydrogen – fuse together, energy is emitted from the new, more stable helium nucleus as gamma radiation.

> **Nuclear fusion:** The combination of two small nuclei to make one large one, with the release of energy as a result

## Fusion at high temperatures

Nuclei are positively charged which means they are repelled by electrostatic forces (see page 37 for more information). At low temperatures the particles are moving slowly which means they do not have enough kinetic energy to overcome this repulsive force. As the temperature increases, average speed and kinetic energy both increase and the nuclei get much closer when they collide. At very close range a special attractive force (called the strong nuclear force) overcomes the electrostatic repulsion and fusion can happen.

The temperature needed to cause fusion is in the millions of kelvin. Nuclear fusion is the energy source of stars, starting with the fusion of hydrogen nuclei to make helium. This happens at the core of the star where both temperature and pressure are high, making fusion more likely. Physicists are starting to understand how to make fusion happen on Earth, but so far it produces less energy than is initially needed to maintain the high temperature and pressure.

> **Exam tip**
>
> Remember that fusion is about what happens between nuclei, not the atoms. The nuclei are strongly positive which is why the repulsion is important even if the atoms are neutral overall.

## Nuclear fission and nuclear fusion

Both nuclear fission and nuclear fusion are sources of energy, which is released because of the mass change that happens when more stable nuclei are formed. Table 7.1 compares the two processes.

**Table 7.1 Comparing fission and fusion**

| Fission | Fusion |
|---|---|
| Smaller nuclei from a large one | Large nucleus from smaller ones |
| Caused by a neutron colliding with a large nucleus | Caused by high temperatures and pressures making small nuclei collide |
| Produces unstable daughter nuclei and neutrons which can cause a chain reaction | Produces stable nuclei such as helium |
| Products carry away kinetic energy | Gamma radiation carries away energy |
| Controlled in a fission reactor and used to generate electricity | Not yet used for the generation of electricity |

## Now test yourself

TESTED ☐

24 Why is alpha emission more like fission than fusion?
25 What is the fuel for the main nuclear fusion process in our Sun?

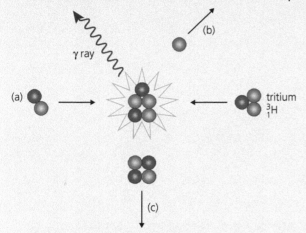

Figure 7.7 **Fusion of two hydrogen isotopes**

26 Identify the particles (a), (b) and (c) in Figure 7.7.

Answers on page 138

**Revision activity**

Being able to explain fission and fusion clearly is very important. Diagrams can be a very good way to check your understanding, so practise labelling and explaining the processes, and using these prompts to write descriptions in clear paragraphs. For the equations, use sketches of the nuclei at each stage to show how the changes happen.

## Summary

- Atoms are made of protons and neutrons (in the nucleus) and electrons (orbiting).
- The atomic or proton number of an element is the number of protons in the nucleus, which have a positive charge.
- The mass, or nucleon number, is the total of the protons and neutrons found in the nucleus.
- Electrons have a tiny mass and a negative charge. For an uncharged atom, the number of protons is equal to the number of electrons. If electrons are gained or lost the atom is called an ion.
- Isotopes have the same number of protons (so are the same element with the same chemical reactions) but different numbers of neutrons (so have different masses).
- Unstable atoms decay and emit one or more kinds of nuclear radiation. This is called radioactivity and the composition of the nucleus is changed in the process.
- An alpha particle is an emitted helium nucleus. These particles are stopped by a few centimetres of air or thin paper and are highly ionising (atomic number −2, mass number −4).
- A beta particle is an electron emitted from the nucleus when a neutron decays into a proton. They are stopped by a few millimetres of aluminium and have medium ionising power (atomic number +1, no change to mass number).
- A gamma ray is a form of electromagnetic radiation emitted after another nuclear decay. They are stopped by lead or thick concrete and are weakly ionising (no change to atomic number or mass number).

- Neutrons are sometimes emitted during nuclear fission or fusion. They are stopped by water or concrete and are weakly ionising (no change to atomic number, mass number −1).
- Nuclear radiation causes harm because of ionisation, which damages living cells and can cause mutations which lead to cancer.
- Gamma radiation is used to kill cancerous cells in the body and to sterilise materials such as surgical instruments. Radioactive tracers are used for diagnosis because the small amount of radiation they emit can be tracked from outside the body.
- The higher the activity of a source, measured in becquerels (Bq), the shorter the half-life, $t_{\frac{1}{2}}$.
- Half-life is the amount of time taken for the activity and the number of unstable atoms to be reduced by half. Each isotope has a consistent value for the half-life, which can range from a fraction of a second to millions of years.
- Nuclear fission of large nuclei such as uranium releases large amounts of energy and neutrons which can continue a chain reaction. This can be controlled in nuclear reactors to generate electricity.
- Nuclear fusion of small nuclei releases large amounts of energy and happens naturally in the Sun, producing larger and more stable nuclei. It only happens at high temperatures because the electrostatic repulsion of positively charged nuclei can only be overcome by fast-moving particles.

## Exam practice

1  A radioactive source and a detector are set up in a school laboratory. Different materials are placed in between the source and detector.

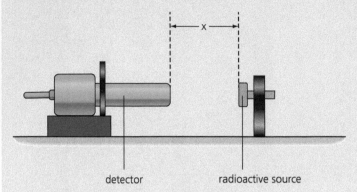

detector         radioactive source

(a) State the name of the detector. [1]

(b) (i) The count rate detected does not change as distance x is increased from 1 cm to 1 m. What type of radiation is **not** emitted? [1]

    A  alpha
    B  beta
    C  gamma
    D  neutrons

   (ii) Explain how you can tell. [1]

(c) Different materials are placed in between the source and detector. The reading is halved when a sheet of aluminium is used but close to zero if a sheet of lead is added. Explain how this shows what is being emitted by the source. [4]

(d) State which material is often used as shielding for neutron radiation. [1]

2  Two isotopes of carbon are $^{12}_{6}C$ and $^{14}_{6}C$.

(a) (i) State how the nuclei are similar. [1]
   (ii) Identify the difference between the nuclei. [1]

(b) $^{14}_{6}C$ is radioactive and decays to nitrogen as shown in the equation. Copy and complete the equation to describe the decay fully. [4]

$$^{14}_{6}C \rightarrow \, ^{?}_{?}N + \, ^{?}_{?}\beta$$

3  Radioactive isotopes have several different uses in medicine.

(a) Explain why a gamma emitter might be used to process surgical equipment. [2]

(b) A different gamma emitter with a short half-life is inhaled by a patient.
   (i) Define half-life. [2]
   (ii) Explain why is it important that an emitter which is taken into the body has a short half-life. [1]

(c) The original activity of a gamma emitter was 200 Bq, and after 18 hours it was measured as 25 Bq. Calculate the half-life. [2]

(d) The element that was used has an isotope with an even shorter half-life, but it is an alpha emitter. Give two reasons why this would not be suitable. [2]

4  A scientist describes the reaction in a fission reaction as follows:

$$^{235}_{92}U + \, ^{1}_{0}n \rightarrow \, ^{92}_{36}Kr + \, ^{141}_{56}Ba + \, ? \, + energy$$

(a) Identify what the question mark in this equation represents. [2]

(b) Explain how a chain reaction is controlled in a fission reactor. You may draw a diagram as part of your explanation. [6]

## Answers and quick quizzes online

ONLINE

# 8 Astrophysics

## Earth's place in the Universe

Our planet, Earth, orbits a star we call the Sun. Earth is part of a solar system that is made up of everything that orbits the Sun as a result of the pull of gravity: planets and moons, dwarf planets, asteroids and comets.

Some planets are mainly solid while others are made up of gas. Lumps of rock too small to be planets are called dwarf planets if they are roughly spherical. Some of the planets in our solar system have one or more moons orbiting them. Asteroids are the rocks that orbit the Sun in between Mars and Jupiter. Comets are rocks that go around the Sun in elongated elliptical orbits, so their distance to the Sun changes over time.

Our Sun is one of several hundred billion stars in a galaxy called the Milky Way. A galaxy is a large group of billions of stars. The Milky Way galaxy is one of many billions of galaxies in the Universe. This is what we mean by the Universe – a large collection of billions of galaxies.

### Now test yourself

TESTED

1 (a) What object do the planets orbit around in our solar system?
  (b) What might orbit around one of these planets?
2 If there are one hundred billion galaxies in the Universe and one hundred billion stars in each galaxy, how many stars are there in the Universe?

Answers on page 138

### Exam tip

Remember, 1 billion = $1 \times 10^9$ or one thousand million.

## Orbits

### Gravitational field strength

REVISED

Every object in the solar system has gravity. The force 'down' for 1 kg of mass towards the centre of each object is called the **gravitational field strength**, g, and it is different for each object. On Earth g is about 10 N/kg. We call the force caused by gravity 'weight'. See page 8 for more information.

Gravity happens because of mass. The larger the mass of an object, the higher the gravitational field strength will be. If two objects with a different radius have the same mass then the smaller one will have a higher value for g at the surface.

**Gravitational field strength:** The strength of gravity on a planet or moon, measured by finding the force down on each kilogram. For some objects, such as asteroids, it may be so small that it is hard to measure.

### The pull of gravity and orbits

REVISED

Close to Earth, gravity causes objects to fall to the surface. Further from the Earth – or further from any large body – gravity causes any objects with mass to have an attractive force between them. This causes smaller masses to orbit larger ones. For example:

● artificial satellites orbit the Earth (and could orbit other planets)
● moons orbit planets and dwarf planets
● planets orbit the Sun.

In each case, the force of attraction causes the smaller mass to continuously change direction and so it orbits the larger mass. Although the orbits are actually elliptical (oval) they are close enough to a circle for most calculations.

## Speed of orbits

REVISED

A satellite in orbit around the Earth is following the path labelled B in Figure 8.1. If the satellite speeds up, the force of attraction is no longer strong enough for it to change direction around the planet and it will fly off into space (path C). If it slows down, the change of direction will bring it closer and closer to the planet until it falls to Earth (path A). A satellite can theoretically orbit a planet at any range, but the closer it is, the faster it must travel to be stable.

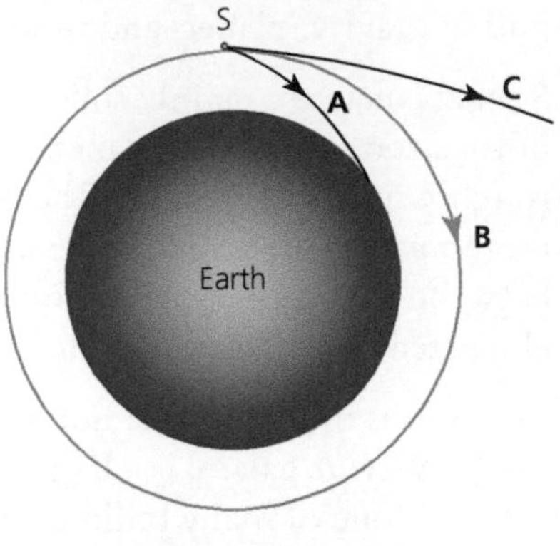

If an orbital path is treated as a circle, the orbital radius is the distance to the centre and the distance travelled is the circumference. The time taken for one complete orbit is called the time period, $T$. Planets orbiting further from the Sun have a lower orbital speed and a larger time period.

$$\text{orbital speed} = \frac{2 \times \pi \times \text{orbital radius}}{\text{time period}}$$

$$v = \frac{2 \times \pi \times r}{T}$$

**Figure 8.1 An artificial satellite, S, is travelling around the Earth.**

orbital speed, $v$, measured in metres per second (m/s)

orbital radius, $r$, measured in metres (m)

time period, $T$, measured in seconds (s)

### Example

How fast is the Earth orbiting if the distance to the Sun is 150 million km on average? Take 1 year as 365 days.

#### Answer

$$v = \frac{2 \times \pi \times r}{T}$$

$$v = \frac{2 \times \pi \times \left(150 \times 10^9\right)}{\left(365 \times 24 \times 60 \times 60\right)}$$

$$v = \frac{3 \times \pi \times 10^{11}}{31\,536\,000}$$

$$v = 29886 \text{ m/s or } 30 \text{ km/s (2 s.f.)}$$

### Exam tip

It can really help to sketch the orbit for a question like this, as you can then label the radius and circumference before attempting the maths.

## Comets

REVISED

Lumps of rock and ice called comets also orbit the Sun. The ice in these comets is melted by the Sun's heat as they get closer and this produces the cometary 'tail'. Each comet is slightly different in time period and composition, but their behaviour is similar. Rather than nearly circular, the orbits of comets are highly elliptical and they are only close to the Sun for a small proportion of their journey. They travel fastest when they are close to the Sun, because gravity causes more acceleration, then slow down as they move further away.

### Revision activity

Sketch the solar system, concentrating on examples of each object rather than the details of names and numbers. Add comparisons of orbital speed and time period to your sketch and check that you can explain the patterns.

# Stellar evolution

## Star temperature and colour

REVISED

Stars emit different kinds of electromagnetic radiation, including visible light. Stars can be described by their brightness, but this is only a measure of how much visible light can be detected here on Earth. Brightness is not the same as the **luminosity**, which is a measure of how much energy is emitted each second across all wavelengths, visible and otherwise. A star which is further away will seem less bright because less of the radiation reaches us. The luminosity is how astrophysicists compare stars, rather than the brightness.

A more useful description of a star is based on colour, which is related to surface temperature. Red or orange stars are the coolest, while blue–white or blue stars are the hottest. Our Sun has a surface temperature of 5800 K, in between these extremes, and is described as yellow–white.

**Luminosity:** The power of the energy emitted by a star, in all directions and across all wavelengths of electromagnetic radiation. This is measured in watts (W).

## Lifecycle of a star

REVISED

The lifecycle of a star, from start to end, occurs over billions of years. What happens at each stage will depend on various factors including the size and mass, because this changes how gravity affects the nuclear fusion process.

The lifecycle starts in space where there is dust and gas, which are individual atoms spread far apart. This collection of dust and gas is called a **nebula** (from a word for cloud), because they look like clouds in the night sky. Gravity causes some of these particles to move together, first slowly, then more quickly, and when a central mass forms it is called a **protostar**. The weight of the outer layers continues to squeeze the atoms closer together and so both pressure and temperature increase (see page 84 for more information). When the nuclei collide quickly enough, nuclear fusion starts, turning hydrogen into helium, and what forms next is described as a **main sequence star**. Our Sun is currently about halfway through its life and has around 5 billion years left as a stable main sequence star.

**Nebula:** A cloud of dust and gas in space which slowly collapses into itself because of gravity

**Protostar:** A ball of gas which contracts due to gravity and heats up due to pressure

**Main sequence star:** A stable gas sphere which has ongoing nuclear fusion, releasing energy as electromagnetic radiation in all directions

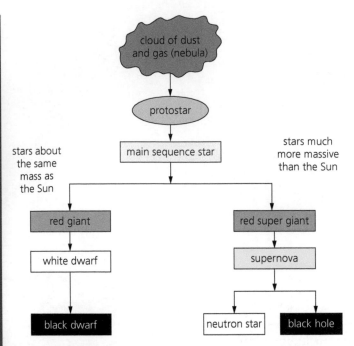

**Figure 8.2 The end stage of a star's lifecycle depends on its mass.**

## Stars about the size of the Sun

Stars like our Sun will eventually run out of the hydrogen atoms that fuel the fusion process. This causes the pressure to drop, so the forces holding the surface away from the core decrease and the star collapses. This causes the core temperature of the star to increase, even though the outer surface is cooler, and the star increases in size again as new fusion processes turn helium into heavier elements such as carbon and oxygen. The star is now a **red giant**. When the helium runs out, the core cools down again and the star shrinks to become a **white dwarf**, which is smaller and has a hot surface. Over time this white dwarf cools and will eventually enter a **black dwarf** stage.

## Stars with a mass much larger than the Sun

If a star has at least 10 times the mass of our Sun, the process is different. As helium is fused to make heavier elements, the star forms a **red super giant**. There is enough mass so that the higher gravity causes more pressure, and so atoms as large as iron can be formed. Eventually, even these huge stars start to run out of atoms for the fusion process. The core will then cool down and the surface collapses, but this happens so fast that the rapid crushing of the core causes the star to explode in a **supernova**. This explosion scatters the outer layers out into space, with atoms larger than iron that were formed during the explosion.

The core of the star is left behind after this supernova but it is now so dense that even atoms cannot exist. Protons and electrons are crushed together to make neutrons, and this **neutron star** may only be a few kilometres across. If the star originally had enough mass, the collapsing nuclei will form a **black hole**, a microscopic point with such strong gravity that nothing nearby, not even light, can escape.

**Red giant**: A star that is fusing helium to make heavier elements up to oxygen

**White dwarf**: A star that is shrinking and is no longer stable as nuclear fusion is stopping

**Black dwarf**: A star that has completely stopped fusing atoms and is no longer emitting heat or light

**Red super giant**: A star that is fusing helium and other nuclei to make atoms as large as iron

**Supernova**: An explosion that sends out the outer layers of a red super giant, leaving behind a dense core

**Neutron star**: A dense core left behind after a supernova, made up of neutrons

**Black hole**: A collapsed mass left behind after the largest stars explode in a supernova, with such a strong gravity even light cannot escape

## Now test yourself

TESTED

6  Why will our Sun never make iron atoms?
7  Why are black holes not visible, even with the most powerful telescopes?
8  What is the difference between a protostar and a main sequence star?

Answers on page 138

# Brightness of stars and absolute magnitude

## Absolute magnitude

REVISED

Historically, astronomers always described stars by their brightness. A first magnitude star was brighter than a second magnitude star, and so on. The problem with this description is that the apparent brightness of a star depends not only on how much light it is emitting, but also on how close it is to us. Scientists now use the idea of **absolute magnitude**, which measures how bright the star would appear at a standard distance. The standard distance chosen is 10 parsecs, or 32.6 light years.

By using a standard distance, comparing the absolute magnitude of two stars is meaningful. A higher value means it is less bright, and the more negative the number the brighter a star would appear at the set distance. As absolute magnitude decreases, the luminosity of the star increases.

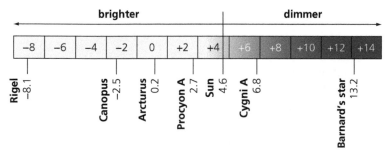

**Figure 8.3** Absolute magnitude allows stars to be compared as if they were at the same range.

> **Exam tip**
>
> You are not expected to know the value of this range for brightness, or how the units are worked out. The important thing to know is that using the same range means the values for different stars are comparable.

> **Absolute magnitude:** A value showing how bright a star would appear at a standard distance from the observer. Stars with a higher value than our Sun (4.6) would appear dimmer, if they were at the same distance. Stars which are very bright have a negative value for the absolute magnitude.

> **Typical mistake**
>
> The scale for absolute magnitude is complex, so make sure you explain clearly that a higher value for magnitude means a less bright object to gain marks. Be particularly careful when comparing values with opposite signs.

## Hertzsprung-Russell diagram

REVISED

As we have seen, the colour of a star gives information about its temperature. This value is plotted on the horizontal axis. The absolute magnitude of a star can be calculated to give information about the luminosity and this is plotted on the vertical axis. This is the origin of the Hertzsprung-Russell (HR) diagram, which has been used for over a century to describe stars in the Universe.

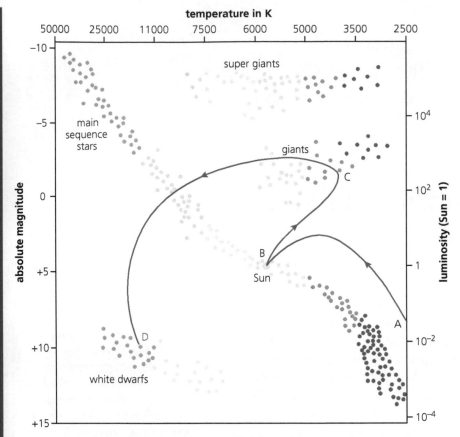

**Figure 8.4** The Hertzsprung-Russell (HR) diagram

If there was no link between luminosity and temperature, we would expect the points to be scattered at random. Instead, different types of stars tend to be in groups. Three of these areas are particularly interesting:

- Main sequence stars form a band or stripe diagonally across the diagram. The brightest stars are also the hottest. Note that all these stars are turning hydrogen into helium by nuclear fusion.
- Stars near the end of their lives, red giants and red super giants, are in two linked groups at the top right of the diagram. These are bright, but at a lower temperature than the equivalent main sequence stars. Nuclear fusion is still happening but larger nuclei are forming.
- Stars in which nuclear fusion is stopping, but without a supernova, are in the bottom left of the diagram. White dwarfs and similar objects are very dim despite the high (and decreasing) temperature.

## The Sun's evolutionary path

REVISED

Our closest star, the Sun, will change in the same way as similar sized stars. Points A to D show how it would appear on the diagram at different stages in its lifecycle:

- in a bright but cool nebula (A)
- currently halfway through 10 billion years as a main sequence star (B)
- getting hotter and slightly dimmer as a red giant with fusion of hydrogen to helium (C)
- collapsing into a white dwarf as fusion slows then stops (D).

> **Exam tip**
>
> The changes in the temperature and behaviour of a star are sometimes called a lifecycle, and sometimes 'stellar evolution'. Stars are not alive, but these terms are a useful way to describe the predictable pattern of changes.

## Now test yourself

TESTED

9  (a)  What happens to the absolute magnitude of a main sequence
        star as it becomes a red super giant?
   (b)  How does its position on the Hertzsprung-Russell diagram
        change?
10  Which is brighter, Procyon A with an absolute magnitude of 2.7, or
    Canopus with an absolute magnitude of −2.5?
11  What kind of value for absolute magnitude would you expect a
    galaxy to have? Explain your answer.
12  Sketch the HR diagram, including labelled axes.

Answers on page 138

### Revision activity

Practice making links
between the descriptions
of a star's luminosity and
temperature, as shown
on the HR diagram, and
the stage of its lifecycle.
Focus on the measurable
and observable changes,
remembering that they are
seen over millions or billions
of years.

# The evolution of the Universe

## The Big Bang Theory

REVISED

The current understanding of the beginning of the Universe is based on
a 'Big Bang' about 13.8 billion years ago. The cause of this explosion is
still disputed, but it is likely to have resulted in highly energetic particles
and gamma rays for the first fractions of a second, which then cooled to
produce protons, neutrons and electrons.

### The first 30 minutes

Within around 30 minutes the temperature from the Big Bang would
have cooled to around $10^8$ K. At this point, some nuclear fusion would
have occurred, producing deuterium and helium nuclei. The temperature
would have been too high for any electrons to orbit nuclei, so no atoms
would have formed and most mass would still be in the form of protons.

### The first billion years

In the first billion years, the Universe continued to expand, which would
have cooled the matter down to a point where atoms could form, after
around 70 000 years. These atoms would originally have been hydrogen
and helium as no more fusion could have happened at first.

Gravity began to cause these atoms to be attracted to each other. This
would have formed clouds or nebulae, which in turn made more fusion
possible as stars and galaxies developed.

### One billion years to now

The most distant objects in the Universe are galaxies that formed a few
billion years after the Big Bang. In comparison, our Sun is relatively
young and formed more than 9 billion years after the Big Bang (about
4.8 billion years ago).

In the billions of years in between the Big Bang and now, stars would
have lived and died, making heavier elements during fusion and
supernova explosions, which in turn produced new stars or planets. Most
of the atoms in your body – everything above helium in the Periodic
Table – were originally made by a star.

## Now test yourself

TESTED

13  Which particle joins with a proton to make a deuterium nucleus?
14  Over billions of years the Universe has expanded and cooled. What had to have happened to the nuclei for atoms to form?

Answers on page 138

# The evidence for the Big Bang Theory

There are two sets of observable data that support the Big Bang Theory. This data does not **prove** that the theory is correct, but currently no other model explains both of these data sets as fully as the Big Bang.

## Cosmic microwave background (CMB) radiation

REVISED

Scientists have found that no matter where a telescope is pointed, it will detect an electromagnetic signal with a wavelength of around 1 mm. This signal is in the microwave band (see page 48 for more information) and corresponds to a background temperature of 2.7 K. This figure matches the temperature predicted for a Universe that has expanded and cooled. As this wavelength is detected from space (cosmic), is between the radio and infrared bands (microwave) and comes from everywhere at once (background), it is called the cosmic microwave background (or CMB).

## Red-shift

REVISED

If the source of a wave is moving towards or away from an observer, the detected wavelength will be different to the emitted wavelength. For sound this difference is called the Doppler effect (see page 46 for more information). This change also happens with electromagnetic waves, but the source has to be moving very quickly to make a measurable difference.

If the source of the electromagnetic wave is moving away, the wavelength increases (and the frequency decreases). Visible light moves towards the red end of the spectrum so it is said to be **red-shifted**. If the source is moving towards the observer, the wavelength decreases. The term blue-shift is sometimes used for this.

Stars do not give out every single wavelength of electromagnetic radiation. There are 'missing wavelengths' of set values, which correspond to the properties of gases in the star. When electromagnetic signals from distant galaxies are detected, the positions of these gaps in the spectrum are different. Since 1929 the evidence obtained has told us:

1  The galaxies must be moving away from us, because the electromagnetic waves are red-shifted.
2  The further away the galaxy, the faster it is moving, because the red-shift is larger.
3  Galaxies are moving away from us in all directions, because this effect is consistent no matter which galaxies are examined.

> **Red-shifted**: An increase of wavelength of a detected electromagnetic wave compared to the source, caused as it moves away from the observer

> **Typical mistake**
>
> The name red-shift can be deceptive; it refers to what will happen for visible light, which is shifted towards the red end of the spectrum. A signal in the red part of the visible spectrum is shifted towards the infrared. The rule to remember is that wavelength always increases for a receding object.

The only way to explain this is that all of the matter in the Universe is spreading out or expanding. At one point, all of the material in the Universe must have been in the same position. When the speeds of the galaxies are compared, we can calculate that all the particles started spreading out nearly 14 billion years ago. This was the Big Bang.

## Calculating the speed and distance of galaxies REVISED

Comparing the amount of red-shift with the reference or emitted wavelength allows the velocity of the galaxy to be calculated:

$$\frac{\text{change in wavelength}}{\text{reference wavelength}} = \frac{\text{velocity of a galaxy}}{\text{speed of light}}$$

$$\frac{\lambda - \lambda_0}{\lambda_0} = \frac{\Delta\lambda}{\lambda_0} = \frac{v}{c}$$

observed wavelength, $\lambda$, measured in metres (m)

reference (emitted) wavelength, $\lambda_0$, measured in metres (m)

change in wavelength, $\Delta\lambda = \lambda - \lambda_0$, measured in metres (m)

velocity of galaxy, $v$, measured in metres per second (m/s)

speed of light in a vacuum, $c$, which is $3 \times 10^8$ m/s

### Example

A reference wavelength of 660 nm is shifted to 690 nm for a distant galaxy. How quickly is it moving away?

Answer

$$\frac{\lambda - \lambda_0}{\lambda_0} = \frac{\Delta\lambda}{\lambda_0} = \frac{v}{c}$$

$$\frac{690 - 660}{660} = \frac{v}{3 \times 10^8}$$

$$v = \frac{3 \times 10^8 \times 30}{660}$$

$$v = \frac{9 \times 10^9}{660}$$

$$v = 1.4 \times 10^7 \text{ m/s (2 s.f.)}$$

### Exam tip

This equation shows why the effect is not easily measurable with moving objects on Earth, as the speed will always be a tiny fraction of the speed of light.

The astronomer Edwin Hubble showed that for closer galaxies, the distance was approximately proportional to the speed. Later work showed that this was true even for more distant galaxies, so working out the speed of a galaxy from its red-shift lets us work out the distance.

### Now test yourself TESTED

15 Why can we not detect red-shift in the light from the Sun?
16 A galaxy is moving away from us at $1.2 \times 10^7$ m/s. Calculate the wavelength change for a reference signal at 550 nm.
17 Explain each part of the term 'cosmic microwave background radiation'.

Answers on page 138

## Summary

- The solar system contains planets, dwarf planets, comets and asteroids orbiting a star, our Sun.
- Galaxies, like our Milky Way, contain billions of stars. The Universe contains many billions of galaxies.
- Gravitational field strength varies depending on the size of the object, and is the cause of moons orbiting planets and planets orbiting stars. Orbital speed is calculated by:

$$v = \frac{2 \times \pi \times r}{T}$$

- Stars are classified by colour which describes the surface temperature from blue-white (above 10 000 K) to red (less than 3500 K).

- Stars change during their lifetime and go through distinct stages.
- The position of stars on the Hertzsprung-Russell diagram shows how luminosity and surface temperature are related.
- Observations of the cosmic microwave background radiation and red-shift of light from galaxies support the theory of the Universe starting 13.8 billion years ago in the Big Bang.
- Measuring red-shift allows the calculation of the speed of a light source by:

$$\frac{\lambda - \lambda_0}{\lambda_0} = \frac{\Delta\lambda}{\lambda_0} = \frac{v}{c}$$

## Exam practice

1 Nuclear fusion is important in several astrophysical processes.
   (a) Explain how deuterium nuclei might have formed in the early Universe. [2]
   (b) Explain what causes nuclear fusion to start when a main sequence star is formed from a protostar. [2]
   (c) Describe how nuclear fusion in a red giant is different to that in a main sequence star like our Sun. [2]
   (d) Explain what happens to the energy produced by nuclear fusion in a star. [2]

2 Phobos is a moon of the planet Mars.
   (a) Define a 'moon'. [2]
   (b) Phobos has an orbital speed of 2.1 km/s and a time period of 460 minutes. Calculate the orbital radius. [3]
   (c) Phobos is small with a mass much lower than that of Mars. Identify which object would have the strongest gravity of the two. Explain your answer. [2]

3 Astronomers observe large stars that are described as red giants and red super giants.
   (a) Explain the difference between a red giant and a red super giant. Include a description of how the end result of stellar evolution will be different for each of them. [6]
   (b) Uranium is naturally present on Earth. Explain what this suggests about how the Earth formed. [3]

4 The cosmic microwave background (CMB) radiation and galactic red-shift are evidence for the Big Bang.
   (a) (i) Define the term cosmic microwave background radiation. [3]
       (ii) Suggest how the wavelength of the CMB radiation will change over the next billion years. [1]
   (b) Red-shift causes: [1]
       A   wavelength to increase, frequency to increase
       B   wavelength to decrease, frequency to increase
       C   wavelength to decrease, frequency to decrease
       D   wavelength to increase, frequency to decrease.
   (c) A galaxy has a speed of $1.5 \times 10^7$ m/s. Calculate the effect on a reference wavelength of 450 nm, given that $c = 3 \times 10^8$ m/s. [3]

**Answers and quick quizzes online**

ONLINE

# Now test yourself answers

## 1 Forces and motion

1. $\text{speed} = \dfrac{\text{distance travelled}}{\text{time taken}}$

   $\text{speed} = \dfrac{1000 - 400}{100 - 70}$

   $\text{speed} = \dfrac{600}{30}$

   $\text{speed} = 20\,\text{m/s}$

2. $\text{average speed} = \dfrac{\text{total distance travelled}}{\text{time taken}}$

   $\text{average speed} = \dfrac{1000}{100}$

   $\text{average speed} = 10\,\text{m/s}$

3. The average speed includes the time when the car was moving more slowly or was stopped.

4. $\text{average speed} = \dfrac{\text{total distance travelled}}{\text{time taken}}$

   $\text{average speed} = \dfrac{10\,000}{3600}$

   $\text{average speed} = 2.8\,\text{m/s (to 1 d.p.)}$

5. (a) $\text{acceleration} = \dfrac{\text{change in velocity}}{\text{time taken}}$

   $\text{acceleration} = \dfrac{+15}{3}$

   $\text{acceleration} = +5\,\text{m/s}^2$

   (b) $\text{acceleration} = \dfrac{\text{change in velocity}}{\text{time taken}}$

   $\text{acceleration} = \dfrac{-12}{6}$

   $\text{acceleration} = -2\,\text{m/s}^2$

6.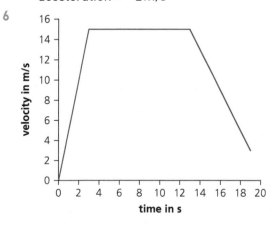

7. $\text{acceleration} = \dfrac{\text{change in velocity}}{\text{time taken}}$

   $\text{time taken} = \dfrac{\text{change in velocity}}{\text{acceleration}}$

   $\text{time taken} = \dfrac{-3}{-2}$

   $\text{time taken} = 1.5\,\text{s}$

8. (a) $v^2 = u^2 + 2 \times a \times s$

   $v^2 - u^2 = 2 \times a \times s$

   $s = \dfrac{\left(v^2 - u^2\right)}{2 \times a}$

   (b) $v^2 = u^2 + 2as$ becomes $v^2 = 2as$ then becomes

   $v = \sqrt{(2as)}$

9. $v^2 = u^2 + 2as$ becomes $v^2 = 2as$ then becomes

   $v = \sqrt{(2as)}$

   $v = \sqrt{(2 \times 2 \times 100)}$

   $v = \sqrt{(400)}$

   $v = 20\,\text{m/s}$

10. (a) $\text{initial speed} = \dfrac{\text{distance}}{\text{time}}$   $\text{final speed} = \dfrac{\text{distance}}{\text{time}}$

    $\text{initial speed} = \dfrac{0.1}{0.1}$   $\text{final speed} = \dfrac{0.1}{0.2}$

    $\text{initial speed} = 1\,\text{m/s}$   $\text{final speed} = 0.5\,\text{m/s}$

    (b) There is no value given for the time between the two speed measurements, so acceleration cannot be calculated. It *is* possible to say it is decelerating, because $v < u$.

11. Thrust should be labelled 900 N and the 700 N arrow should be labelled drag or air resistance.

12. The ball is pushed which increases the speed. It will then slow down because of friction and/or air resistance. The wall exerts a force to change the shape, compressing it temporarily. When it returns to the original shape, this exerts a force on the wall to change the direction of motion.

13. (a) 300 N forwards or to the right, causes acceleration (forwards or to the right)

    (b) Balanced or zero force, no effect on movement

    (c) 2000 N backwards or to the left, causes deceleration (or acceleration backwards)

14. If the engine stops working the force acting forwards becomes zero. There are still forces acting to resist motion, for example air resistance. The forces are no longer in balance so there will be a change in speed, in this case deceleration.

15. $F = m \times a$

    $a = \dfrac{F}{m}$

    $a = \dfrac{40\,000}{200\,000}$

    $a = 0.2\,\text{m/s}^2$

16 (a) $W = m \times g$
   $W = 900 \times 10$
   $W = 9000\,N$ or $9\,kN$

   (c) $W = m \times g$
   $W = 900 \times 4$
   $W = 3600\,N$ or $3.6\,kN$

   (b) Mass is the same, so $900\,kg$

17 The skydiver with an open parachute has the lowest terminal velocity. This is important because the speed needs to be low enough that landing is survivable.

18 thinking distance = speed × reaction time

   thinking distance = $20 \times 0.9$

   thinking distance = $18\,m$

19 The total stopping distance is increased. This is because ice on the road reduces friction between the road and the tyres which increases braking distance.

20 From $5\,cm$ to $7\,cm$ is an extension of $2\,cm$. If the force is doubled, so is the extension, so $4\,cm$.

21 (a) $F = k \times e$
   $k = \dfrac{F}{e}$
   $k = \dfrac{4}{0.02}$
   $k = 200\,N/m$

   (b) $F = k \times e$
   $e = \dfrac{F}{k}$
   $e = \dfrac{1000}{200}$
   $e = 5\,m$

   (c) In practice, the spring would probably undergo plastic (non-proportional) deformation and break long before this point.

22 Elastic deformation is temporary and the object will return to the original size and shape once the force is removed. An object which has been plastically deformed will not.

23 $p = m \times v$
   $p = 0.06 \times 20$
   $p = 1.2\,kg\,m/s$

24 Momentum is a vector quantity as it has direction as well as magnitude (size).

25 (a) $\Delta p = m \times \Delta v$
   $\Delta p = 65 \times 3$
   $\Delta p = +195\,kg\,m/s$

   (b) There is no time given for the change in velocity.

26 $F = \dfrac{\Delta mv}{t}$
   $F = \dfrac{80 \times 25}{0.4}$
   $F = 5000\,N$ or $5\,kN$

27 All momentum is transferred to the first from the second trolley:
   $m_1 \times v_1 = m_2 \times v_2$
   $v_1 = \dfrac{m_2 \times v_2}{m_1}$
   $v_1 = \dfrac{1 \times 0.8}{2}$
   $v_1 = 0.4\,m/s$

28 $m_1 \times v_1 = m_2 \times v_2$
   $v_2 = \dfrac{m_1 \times v_1}{m_2}$
   $v_2 = \dfrac{45 \times 3}{50}$
   $v_2 = 2.7\,m/s$

29 $60\,N$ (in the opposite direction)

30 $M = F \times d$
   $d = \dfrac{M}{F}$
   $d = \dfrac{750}{300}$
   $d = 2.5\,m$

31 They are incorrect. The maximum load will be halved when the distance is doubled (to 32.5 tonnes). It is inversely, not directly, proportional, because it is the turning moment which matters.

32 The newton metre (Nm)

33 In equilibrium so clockwise and anti-clockwise turning moments must be equal:
   $F_1 \times d_1 = F_2 \times d_2$
   $F_2 = \dfrac{F_1 \times d_1}{d_2}$
   $F_2 = \dfrac{4000 \times 12}{2}$
   $F_2 = 24000\,N$ or $24\,kN$

34 Half each so $7.5\,kN$

## 2 Electricity

1 (a) $3 \times 1.5 = 4.5\,V$    (b) The bulb would be brighter.

2 There could be a spark and you might receive an electric shock.

3 Voltage

4 $250\,mA = 0.25\,A$

5 Any three of: earth connection, fuse, cord grip, plastic insulaton

6 Greater numbers of cycles

7 Any two of: gives a more precise value; it's faster at cutting current; faster to replace easy to reset once the problem is solved and the current is back to a safe level; can be set to any value, not just common ones such as a 13 A fuse.

8 Current, $I$, is measured in amperes or amps (A)

9 (a) $P = I \times V$
   $P = 0.5 \times 12$
   $P = 6\,W$

   (b) $E = P \times t$
   $E = 6 \times 300$
   $E = 1800\,J$ or $1.8\,kJ$

10 $P = \dfrac{E}{t}$

$P = \dfrac{400}{60}$

$P = 6.7\,W$ (to 1 d.p.)

11 $I = \dfrac{P}{V}$

$I = \dfrac{1800}{230}$

$I = 7.8\,A$ (to 1 d.p.)

12 (a) (Open) switch      (b) Ammeter

   (c) Variable resistor    (d) Lamp

13 (a) Current decreases    (b) Current decreases

   (c) Current increases

14 The other lamp goes out if one breaks.

15 An ammeter is connected in series, as part of the main loop. The two connections of the voltmeter are placed either side of the component being measured, in parallel.

16 Resistance, $R$, measured in ohms ($\Omega$)

17 $R = \dfrac{V}{I}$

$R = \dfrac{12}{0.4}$

$R = 30\,\Omega$

18 (a) Circuit symbol for a filament lamp

   (b) Horizontal axis labelled voltage/$V$ and vertical axis labelled current/$I$. Line through origin, steep at low voltages then flattening out as voltage increases.

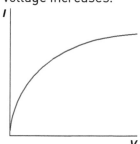

19 $R = 4 + 8 + 6 = 18\,\Omega$

20 (a) Charge, $Q$, is measured in coulombs (C).

   (b) A volt means one joule is transferred for each coulomb of charge.

21 Electrons, which are negatively charged

22 (a) $Q = I \times t$       (b) $E = Q \times V$

     $Q = 0.043 \times 120$      $E = 5.16 \times 2.4$

     $Q = 5.16\,C$           $E = 12.4\,J$ (to 1 d.p.)

23 $Q = I \times t$

$I = \dfrac{Q}{t}$

$I = \dfrac{36}{60}$

$I = 0.6\,A$

24 (a) Parallel

   (b) If the lights were in series, they could either be all on or all off, not a mixture.

25 (a) 3 A                (b) $2 + 2 + 2 = 6\,A$

   (c) $2 + 1.5 = 3.5\,A$

26 (a) 12 V             (b) 12 V

27 (a) 1.5 V           (b) 3 V

   (c) Current through the 15 Ω resistor:

     $I = \dfrac{V}{R}$

     $I = \dfrac{3}{15}$

     $I = 0.2\,A$

   (d) Resistance of the lamp:

     $R = \dfrac{V}{I}$

     $R = \dfrac{3}{0.25}$

     $R = 12\,\Omega$

   (e) Ammeter reading = 0.2 + 0.25 = 0.45 A

28 (a) Negative

   (b) The sign is the same

29 (a) The electrons are free to move even though the atoms themselves are fixed. It is the movement of electrons that makes up the current in a metal wire.

   (b) In plastic and other insulators the electrons are not able to move away from the nuclei, so no current can flow.

30 As the fuel is pumped on to the plane, friction makes it charged. If this charge builds up it could cause a spark and set the fuel/plane on fire. The cable allows the charge to flow back so it is balanced and there is no spark.

31 (a) The particles will all be positivey charged, so they will repel each other and spread out.

   (b) A negative charge so they attract the soot

# 3 Waves

1 (a) Hertz (Hz)

   (b) $T = \dfrac{1}{f}$

     $f = \dfrac{1}{T}$

     $f = \dfrac{1}{0.02}$

     $f = 50\,Hz$

2 A longitudinal wave

3 $v = f \times \lambda$

$$f = \frac{v}{\lambda}$$

$$f = \frac{330}{0.2}$$

$$f = 1650\,Hz$$

4 When a wave meets a surface, the angles are measured from the normal to that surface.

5 28 cm/8 peaks = 3.5 cm

6 The pitch will be high as it approaches then get lower (deeper) as it moves away.

7 (a) Wavelength has decreased/is less

(b) Refraction

8 (a) Radio, micro, IR, visible

(b) UV, X-ray, gamma    (c) IR, visible

9 (a) 350 nm = 3.5 × 10⁻⁷ m

(b) $f = \frac{c}{\lambda}$

$$f = \frac{3 \times 10^8}{3.5 \times 10^{-7}}$$

$$f = 8.6 \times 10^{14}\,Hz\,(to\,2\,s.f.)$$

10 Blue

11 Sunburn is caused by ultraviolet (UV), which is part of sunlight all year round. UV light reflects off the snow making skiers more susceptible to sunburn.

12

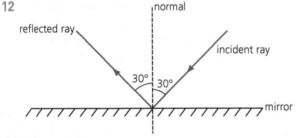

13 90° or right angle

14 85°

15 Less

16 Ray boxes get very hot in use so be careful when moving them to avoid burns.

17 $n = \frac{\sin i}{\sin r}$

$$\sin r = \frac{\sin i}{n}$$

$$\sin r = \frac{\sin 30}{1.5}$$

$$\sin r = \frac{0.5}{1.5}$$

$$r = \sin^{-1} 0.33$$

$$r = 19.5°\,(1\,d.p.)$$

18 refractive index = $\frac{\text{speed of light in air}}{\text{speed of light in the material}}$

speed of light in the material = $\frac{\text{speed of light in air}}{\text{refractive index}}$

speed of light in the material = $\frac{3 \times 10^8}{2.42}$

speed of light in the material = 1.24 × 10⁸ m/s

19 (a) The critical angle is the lowest angle of incidence when all light is reflected, none refracted.

(b) 42°

20 $\sin c = \frac{1}{n}$

$$n = \frac{1}{\sin c}$$

$$n = \frac{1}{\sin 49}$$

$$n = 1.33$$

21 Light is much more likely to be reflected than refracted through the surfaces, which is part of the reason why diamonds sparkle.

22 The light crosses each boundary at a right angle. Refraction only happens when light enters or leaves a material at an angle to the normal.

23 (a) $v = f \times \lambda$    (b) 40 × 60 = 2400 times in a minute

$$\lambda = \frac{v}{f}$$

$$\lambda = \frac{330}{40}$$

$$\lambda = 8.25\,m$$

24 Push and pull on the end to cause compressions and decompressions (rarefactions)

25 They said frequency when they meant amplitude (for loudness).

26 The x-gain (horizontal)

27 (a) Higher peaks/lower troughs

(b) Narrower waveforms, more visible at one time

28 $T = \frac{1}{f}$

$$T = \frac{1}{600}$$

$$T = 1.67\,ms$$

So each wave form will be one third of a centimetre.

## 4 Energy resources and energy transfer

1 (a) Gravitational store    (b) Thermal store

(c) Elastic store    (d) Kinetic store

2 (a) 28 000 J    (b) 4 300 000 J

(c) 910 J    (d) 76 500 000 J

3 efficiency = $\frac{\text{useful energy output}}{\text{total energy input}}$

4

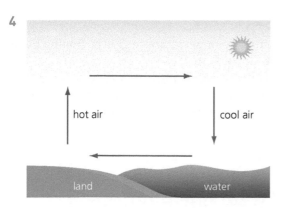

hot air     cool air

land     water

5 The rate of transfer of energy depends on several factors, one of which is the temperature difference. On a cold day the difference in temperature between inside and outside is greater so the transfer will be faster.

6 Metals like copper have free electrons which make them good thermal conductors.

7 The marble will fall off before the one on the shiny surface but after the one on the black surface (between 34 and 53 seconds, based on the sample results given). This is because the grey surface is a better absorber than the shiny one but not as good as the black one.

8 Infrared or IR

9 The snow under the black sheet will melt faster than the snow under the white sheet. This is because the black sheet absorbs more of the thermal radiation on it so gets warmer.

10 To make them better emitters (and so cool down the contents)

11 (a) $\text{distance} = \dfrac{\text{work done}}{\text{force}}$ or $d = \dfrac{W}{F}$

  (b) $\text{force} = \dfrac{\text{work done}}{\text{distance}}$ or $F = \dfrac{W}{d}$

12 $F = \dfrac{W}{d}$

$F = \dfrac{80}{4}$

$F = 20\,\text{N}$

13 $d = \dfrac{W}{F}$

$d = \dfrac{240}{32}$

$d = 7.5\,\text{m}$

14 (a) Energy is transferred electrically (by a current) from the chemical store of the battery to the thermal store of the filament.

  (b) Energy is transferred by conduction and by both visible and IR radiation from the thermal store of the filament to the thermal store of the room.

15 $\text{KE} = \dfrac{1}{2} \times m \times v^2$

$\text{KE} = \dfrac{1}{2} \times 0.06 \times (20)^2$

$\text{KE} = \dfrac{1}{2} \times 0.06 \times 400$

$\text{KE} = 12\,\text{J}$

16 $\text{GPE} = m \times g \times h$

$\text{GPE} = 75 \times 10 \times 12$

$\text{GPE} = 9000\,\text{J or } 9\,\text{kJ}$

17 $P = \dfrac{W}{t}$

$t = \dfrac{W}{P}$

$t = \dfrac{m \times g \times h}{P}$

$t = \dfrac{10 \times 10 \times 8}{40}$

$t = \dfrac{800}{40}$

$t = 20\,\text{s}$

18 $m \times g \times h = \dfrac{1}{2} \times m \times v^2$

$v^2 = 2 \times g \times h$

$v^2 = 2 \times 10 \times 12$

$v^2 = 240$

$v = \sqrt{240}$

$v = 15.5\,\text{m/s (to 1 d.p.)}$

NB: the mass is irrelevant as it cancels out when you rearrange.

19 $P = \dfrac{W}{t}$

$W = P \times t$

$W = 2700 \times (5 \times 60)$

$W = 2700 \times 300$

$W = 810000\,\text{J or } 810\,\text{kJ or } 0.81\,\text{MJ}$

20 There are no fuel costs for solar power but the installation costs of the solar cells are high.

21 Erratic supply, noise and appearance

22 The fuel used in nuclear power stations will run out eventually, even though there is currently no shortage.

## 5 Solids, liquids and gases

1 Gravel is denser so an equal volume has more mass.

2 $\rho = \dfrac{m}{V}$

$m = \rho \times V$

$m = 8000 \times (0.06 \times 0.1 \times 0.06)$

$m = 8000 \times 0.00036$

$m = 2.88\,\text{kg}$

3 Ethanol: $m = \rho \times V$

$m = 0.789 \times 40$

$m = 31.56\,g$

Water: $m = \rho \times V$

$m = 1 \times 60$

$m = 60\,g$

Total mass is 91.56 g (or 92 g to 2 s.f.)

4 (a) $F = P \times A$

(b) $F = P \times A$

$F = 400\,000 \times 0.0012$

$F = 480\,N$

5 The area of the fingertips is smaller than the area of the palm, so the same force (weight) causes a larger pressure and that makes the press-up harder.

6 Polar bears and camels walk on soft ground (snow and sand, respectively). The large area of their feet makes it less likely they will sink in.

7 The pressure is greater at the bottom so the force acting to push the water out of the hole is larger.

8 The pressure is the same on all sides so we do not notice any overall effect.

9 (a) $P = h \times \rho \times g$

$h = \dfrac{P}{\rho \times g}$

$h = \dfrac{100\,000}{1025 \times 10}$

$h = 9.8\,m$ (2 s.f.)

(b) The total pressure must also include the air pressure, which has an average sea-level value of 100 kPa.

10 (a) Freezing (or solidifying)

(b) Condensing or condensation

(c) Deposition

11 The flat line would be at 0 °C because this is where the ice melts (i.e. changes state). While it is melting there is no temperature change.

12 The particles in a solid vibrate. This means they are moving back and forth around a fixed point.

13 The gaps become much smaller.

14 Copper

15 $\Delta Q = m \times c \times \Delta T$

$\Delta Q = 0.15 \times 4200 \times 40$

$\Delta Q = 25\,200\,J$ or 25.2 kJ

16 (a) The measured temperature change would be smaller

(b) The calculated specific heat capacity would be greater (and less accurate)

17 The pressure of the gas increases when it is warmer so there will be a greater force on the inside of the tyre.

18 (a) $273 - 196 = 77\,K$

(b) Melting point of water/ice: 273 K. Boiling point of water/steam: 373 K

19 The gas that is left is now at a lower pressure in the same volume. The particles have less kinetic energy so the temperature is lower.

20 Volume

# 6 Magnetism and electromagnetism

1 (a) Attract     (b) No effect     (c) No effect

(d) Attract south pole, repel north pole

2 (a)

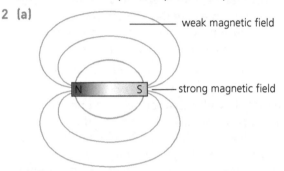

(b) The lines would be closer together

3 (a)

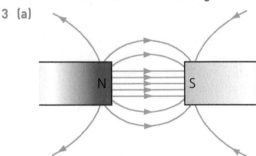

(b) The poles must be opposites.

4 A permanent magnet always has a magnetic field around it; an induced magnet only has a magnetic field when it is within another field or touching another magnet.

5 The two pins have been magnetised and their induced poles are the same, so they repel.

6 A magnetically soft material is used to make an electromagnet, so it can be turned on and off.

7 At the ends (or in the centre)

8 Arrow on a line: shows the direction of conventional current, from positive to negative

Circle with a cross: shows the current going into the paper

Circle with a dot: shows the current coming out of the paper

9 The right-hand grip rule (thumb pointed down to match the arrows for the current) shows that the field lines will point clockwise.

10 (a) More turns of wire in the coil, or add a soft iron core

(b) Test how many paperclips or pins it can hold, or the maximum distance at which it affects a compass

11 It will spin faster.

12 **F**irst finger for the magnetic **F**ield, se**C**ond finger for **C**urrent, Thu**M**b for Thrust/**M**otion

13 From east/west *or* at right angles to north/south

14 The opposite force acts on the permanent magnet, but because this is (probably) much, heavier the motion is much smaller.

15 Nothing; when the coils are vertical there is no force.

16 The motor turning faster might have more turns on the coil, so the force is greater. (It could also be because of less friction, but the question specifies the motor effect.)

17 The alternating current will have a greater value but a lower frequency.

18 The same, 0.24 V. Do not accept −0.24 V.

19 (a) Higher voltage

(b) Same voltage, opposite direction or sign

(c) Lower voltage

20 There will be no induced voltage because the solenoid is not in a changing magnetic field.

21 The changing direction of the magnetic field is what makes the induced voltage alternate. If a split-ring commutator or similar was used, the output would be a direct current.

22 The frequency of the a.c. will also be doubled.

23 The largest voltage is induced when the coil is horizontal, because this is when the wires on each side are moving vertically the fastest.

24 $\dfrac{V_P}{V_s} = \dfrac{N_P}{N_s}$

$N_P = N_s \times \dfrac{V_P}{V_s}$

$N_P = 5000 \times \dfrac{11\,500}{230}$

$N_P = 5000 \times 50$

$N_P = 250\,000$ turns

25 $V_P \times I_P = V_s \times I_s$

$V_s = \dfrac{V_P \times I_P}{I_s}$

$V_s = \dfrac{230 \times 0.2}{2}$

$V_s = \dfrac{46}{2}$

$V_s = 23\,V$

26 There is only an induced voltage on the secondary coil when the magnetic field is changing. This only happens when the primary current is changing, so an alternating supply is needed.

## 7 Radioactivity and particles

1 (a) 3 protons, 4 neutrons     (b) 3     (c) 7

2 Ions have lost or gained electrons in the orbits around the nucleus. Isotopes are atoms with a different number of neutrons but the same number of protons.

3 It has 2 more neutrons and/or has a larger mass.

4 An alpha particle is made up of 2 protons and 2 neutrons, which is the same as a helium nucleus.

5

| | Emitted from nucleus | Change in mass number | Change in atomic number |
|---|---|---|---|
| Alpha | 2p + 2n | −4 | −2 |
| Beta | Electron | 0 | +1 |
| Gamma | EM wave | 0 | 0 |
| Neutron | Neutron | 0 | −1 |

6 (a) All of them: alpha, beta, gamma, neutron

(b) Gamma, neutron     (c) Alpha, beta

7 $^{3}_{1}H \rightarrow {}^{0}_{-1}\beta + {}^{3}_{2}He$

8 Beta

9 Airline pilots will have a higher exposure to cosmic rays, coal miners might be exposed to more radioactive rocks and/or radon.

10 Alpha radiation can't get through the skin but if swallowed or inhaled it can cause a lot of damage.

11 Lead

12 The sample with a short half-life would have a new activity of around 500 Bq. The sample with the long half-life has probably not changed.

13 Small

14 Activity includes emissions which are not detected in the count rate.

15 The digestive system

16 A beta source; no matter how thick the aluminium, it would stop all alpha particles and no gamma rays. Only the amount of detected beta radiation would change depending on the thickness of the aluminium.

17 The human body would absorb all the alpha radiation before it could be detected outside the body. The solution would be to test around the mouth, and hope for a negative result.

18 (a) Alpha is less dangerous than gamma because it is blocked by the skin and has a short range in air.

(b) Alpha is more dangerous than gamma because it causes direct damage to cells and is strongly ionising, which means lots of ions to damage DNA.

19 Lead or concrete

20 If an isotope has a high activity this means the unstable atoms are decaying quickly. This is the opposite to a long half-life, which is when the atoms are more stable.

21 Both isotopes have 92 protons. U-235 nuclei have 143 neutrons and U-238 nuclei have 146 neutrons.

22 Nuclear store (of uranium atom) to kinetic store (of fission products) to thermal store (of graphite and gas) to thermal store (of water/steam) to kinetic store (of turbines) then transferred by electrical pathway.

23 (a) Boron is used in the control rods to absorb neutrons and stop the chain reaction.

(b) Uranium is used in the fuel rods to provide the nuclei for the fission reaction.

(c) Concrete is used as shielding to keep workers safe, because it contains water that absorbs neutrons.

(d) Graphite is used as a moderator to slow down the fast neutrons from one fission reaction so they can be absorbed by another uranium nucleus for the chain reaction.

24 Because smaller nuclei are produced from larger ones

25 (Isotopes of) hydrogen

26 (a) Deuterium (or hydrogen-2)

(b) Neutron     (c) Helium

# 8 Astrophysics

1 (a) The Sun     (b) Moons or satellites

2 A minimum of: $1 \times 10^{22}$

3 Less

4 Neptune

5 $v = \dfrac{2 \times \pi \times r}{T}$

$v = \dfrac{2 \times \pi \times 414 \times 10^9}{4.6 \times (365 \times 24 \times 60 \times 60)}$

$v = \dfrac{8.28 \times \pi \times 10^{11}}{145\,065\,600}$

$v = 17\,931\,\text{m/s}$ or $18\,\text{km/s}$ (2 s.f.)

6 Iron atoms are only made in red super giants, and our Sun does not have enough mass to go through this stage.

7 Light cannot escape from a black hole, which is why they're described as 'black'.

8 In a protostar the pressure and temperature at the core are increasing, but not enough for fusion to start. Once it does, the object is a (main sequence) star.

9 (a) The absolute magnitude of a main sequence star increases when it becomes a red super giant.

(b) The position on the Hertzsprung-Russell diagram moves up and right, diagonally.

10 Canopus (absolute magnitude of −2.5) is brighter (the value might be less but negative magnitudes are brighter).

11 A galaxy would have a strongly negative absolute magnitude, because this is calculated as if it were relatively close.

12

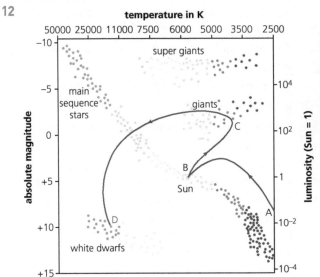

13 Neutron

14 Electron(s) in orbit

15 The Sun is not moving away from us.

16 $\dfrac{\Delta\lambda}{\lambda_0} = \dfrac{v}{c}$

$\Delta\lambda = \dfrac{v}{c} \times \lambda_0$

$\Delta\lambda = \dfrac{1.2 \times 10^7}{3 \times 10^8} \times 550$

$\Delta\lambda = \dfrac{1.2 \times 55}{3}$

$\Delta\lambda = 22\,\text{nm}$

17 Cosmic: from space

Microwave: around 1 mm wavelength, in the microwave band

Background: from all directions equally

Radiation: an EM wave